T0205419

Springer Theses

Recognizing Outstanding Ph.D. Research

Aims and Scope

The series "Springer Theses" brings together a selection of the very best Ph.D. theses from around the world and across the physical sciences. Nominated and endorsed by two recognized specialists, each published volume has been selected for its scientific excellence and the high impact of its contents for the pertinent field of research. For greater accessibility to non-specialists, the published versions include an extended introduction, as well as a foreword by the student's supervisor explaining the special relevance of the work for the field. As a whole, the series will provide a valuable resource both for newcomers to the research fields described, and for other scientists seeking detailed background information on special questions. Finally, it provides an accredited documentation of the valuable contributions made by today's younger generation of scientists.

Theses are accepted into the series by invited nomination only and must fulfill all of the following criteria

- They must be written in good English.
- The topic should fall within the confines of Chemistry, Physics, Earth Sciences, Engineering and related interdisciplinary fields such as Materials, Nanoscience, Chemical Engineering, Complex Systems and Biophysics.
- The work reported in the thesis must represent a significant scientific advance.
- If the thesis includes previously published material, permission to reproduce this must be gained from the respective copyright holder.
- They must have been examined and passed during the 12 months prior to nomination.
- Each thesis should include a foreword by the supervisor outlining the significance of its content.
- The theses should have a clearly defined structure including an introduction accessible to scientists not expert in that particular field.

More information about this series at http://www.springer.com/series/8790

Gregor Posnjak

Topological Formations
in Chiral Nematic Droplets

Doctoral Thesis accepted by
the University of Ljubljana, Ljubljana, Slovenia

 Springer

Author
Dr. Gregor Posnjak
Department of Condensed Matter Physics
Jožef Stefan Institute
Ljubljana, Slovenia

Supervisor
Prof. Dr. Igor Muševič
Faculty of Mathematics and Physics
University of Ljubljana
Ljubljana, Slovenia

ISSN 2190-5053   ISSN 2190-5061 (electronic)
Springer Theses
ISBN 978-3-030-07475-3   ISBN 978-3-319-98261-8   (eBook)
https://doi.org/10.1007/978-3-319-98261-8

This Springer imprint is published by the registered company Springer Nature Switzerland AG
The registered company address is: Gewerbestrasse 11, 6330 Cham, Switzerland

Supervisor's Foreword

In this thesis, Gregor Posnjak resolves the long-standing mystery of the internal director structure of chiral nematic droplets, which has been studied both experimentally and theoretically since the 1970s. Due to spherical confinement of the liquid crystal and the spontaneous twisting of the liquid crystal, a huge variety of different topological states were predicted, including knotted and linked ones. Until recently, the droplets were studied only with polarised optical microscopy, and from these images, it was very difficult to reconstruct the three-dimensional structure of the optically birefringent structures with a number of topological defects. Gregor Posnjak used all the advantages of fluorescent confocal polarised microscopy (FCPM) to reconstruct the full 3D director structure inside the chiral nematic droplets. He used fluorescent molecules that locally align with the nematic molecules to measure the director orientation by taking several consecutive 3D confocal scans at different polarisations. These 3D images of fluorescence were then used as the input for a numerical simulated annealing algorithm that enables the full reconstruction of the 3D director field inside the droplet.

The FCPM and the experimental details are explained in depth in Chaps. 4–6, together with a newly developed simulated annealing algorithm. This algorithm makes possible the automated determination of the direction of the tilt of fluorescent molecules with respect to the focal plane of the microscope and is crucial for a reliable reconstruction of the director. A significant part of the thesis is devoted to testing the reconstruction algorithm on droplets with well-known structures. These methods are then used in Chap. 7 to explore the complex 3D structures in the chiral nematic liquid crystal droplets with perpendicular surface anchoring. As many as 24 distinct topological states are identified and presented in detail, including layered structures of different symmetries and states with multiple topological point defects of unit charge separated by localised chiral structures. The thesis reports on the first observation of topological point defects with higher topological charges $q = -2$ and $q = -3$ that are only stable in tightly confined chiral nematics.

This thesis makes for excellent, in-depth reading, not only for specialists in liquid crystals, but also for a much broader audience, including readers interested in experimental topology and imaging techniques.

Ljubljana, Slovenia   Prof. Dr. Igor Muševič
July 2017

Acknowledgments

Mentorju prof. Igorju Muševiču se zahvaljujem za zanimiv in zahteven raziskovalni projekt, za pomoč in vzpodbude ter za strokovno vodenje pri raziskavah.

Izredno hvaležen sem Simonu Čoparju za vso pomoč, razlage in razprave. Brez njegovega sodelovanja bi to delo verjetno imelo zelo drugačno podobo.

Izpostavil bi tudi pomoč Davida Seča in Žiga Osolina. Žigova pomoč pri algoritmu simuliranega ohlajanja je bila bistvena za uspeh tega doktorskega dela, numerične strukture kapljic, ki jih je priskrbel David, pa so mi izjemno pomagale pri razvoju in testiranju postopka rekonstrukcije.

Zahvaljujem se tudi vsem sodelavcem na IJS: Mihu Škarabotu, Giorgiotu Mirriju, Andriyu Nychu, Matjažu Humarju, Janji Milivojevič, Uliani Ognysta, Maruši Mur, Venkata Suba Rao Jampaniju, Maryam Nikkhou ter Urošu Jagodiču. Med mojim delom so mi bili vedno v pomoč z nasveti in razlagami, obenem pa so bili dobra družba, zaradi katere je delo vedno potekalo v prijetni atmosferi.

Prof. Žumru, Mihu Ravniku in kolegom Urbanu Muru, Žigu Kosu, Anji Bregar in Juretu Aplincu se zahvaljujem za sodelovanje, razprave ter diskusije.

Hvala tudi drugim sodelavcem z inštituta za vso pomoč, ter kolegom, ki so mi pomagali ter mi stali ob strani, še posebej Anžetu, Drejcu in Anji.

Zahvaljujem se tudi svoji družini, še posebej Neži, ki mi ves čas stoji ob strani in me podpira v mojih življenskih izbirah.

Abstract

This thesis deals with director formations in chiral nematic droplets with homeotropic anchoring. To develop a suitable tool for the exploration of these structures, we derive the angular dependence of fluorescence intensity in fluorescent confocal polarising microscopy (FCPM) and discuss which information can be extracted by it from experimental data. We find that FCPM can characterise the orientation of nematic director, with the exception of the sign of the component oriented along the optical axis. We develop a simulating annealing algorithm which determines the most probable structure of the director field from the experimental data by finding the configuration of the missing signs which minimises the elastic energy. The procedure is tested on model and experimental data. We use the FCPM method, augmented with the simulating annealing algorithm to reconstruct director fields from experimental FCPM data of chiral nematic droplets. We characterise 24 types of structures, which can be roughly divided into structures with cholesteric layers, and structures with multiple topological point defects, which are separated by localised chiral structures, dubbed cholesteric bubbles. The bulk volume of the layered structures is similar to structures which were numerically predicted to form in chiral nematic droplets with planar anchoring. In droplets with multiple cholesteric bubbles, the point defects organise in string-like constellations of defects with alternating sign of topological charge. At specific relative chirality ranges, the cholesteric bubbles stabilise higher-charge topological point defects. Two types of such defects are found, with topological charges $q = -2$ and $q = -3$. The point defects can be arranged in topological molecules where a single point defect is substituted with several defects with equivalent total topological charge. The formation of strings and topological molecules is explained by constructing them from simpler droplets to which topologically neutral sets of point defects and cholesteric bubbles are added.

Keywords Liquid crystal droplets · Topological defects · Skyrmions
Fluorescent confocal polarising microscopy · Confinement
Chiral nematic liquid crystals · Cholesterics

Contents

Chapter 1
Introduction

Liquid crystals (LC) are materials with anisotropic physical properties owning to the partial order which arises in them because of the anisotropy of their building blocks [1]. The headless vector field called director, which describes their average orientation, has proven to be a fruitful playground for various realisations of topological concepts because of the ease of manipulation of the liquid crystals and of examination of the structures they form on the micrometre scale. Liquid crystals can be manipulated by temperature, density, electric and magnetic fields and focused laser beams but even just the control of boundary conditions in a confined space can allow for ample variation of stable configurations.

In cases where chirality is introduced into the system, the orientation of the building blocks can rotate in space in a helical fashion. Because of the helical twisting of the director field, the ground state of a chiral nematic is a periodic structure where the director rotates in a plane as you traverse the sample in a direction, normal to this plane. Chiral nematics therefore have an additional degree of freedom compared to non-chiral nematics—the distance over which the director completes a full turn, called the pitch.

If a chiral nematic LC is confined to a small volume where the boundary conditions on the surface of the volume do not match the chiral ordering, the helical structure is strongly frustrated. The interaction between confinement and spontaneous twisting of chiral nematics can give rise to a plethora of topologically rich structures with different elastic energies, but separated by high enough energy barriers for several of them to be stable. These metastable structures can be switched between by melting the LC to isotropic phase and then cooling it back to the LC phase. The structures can also be continuously controlled by either changing the pitch or the anchoring on the surface. An example of this metastability are the skyrmion-like structures, which form in thin films with perpendicular boundary conditions on two parallel sides [2–5]. In this example, one of the possible metastable structures is the unwound uniform director field. The other possible structures maximise the volume of the twisted liquid crystal, while still meeting the boundary conditions. The main parameter which dictates the stability of different structures is the confinement ratio, defined as the

© Springer Nature Switzerland AG 2018
G. Posnjak, *Topological Formations in Chiral Nematic Droplets*,
Springer Theses, https://doi.org/10.1007/978-3-319-98261-8_1

ratio of the distance between the plates with perpendicular boundary conditions and the chiral pitch of the LC.

A similar richness of different metastable structures was predicted for chiral nematic LCs, confined in spherical volumes [6, 7]. If the boundary condition for the director on the surface of the volume is tangential, this lends flexibility to the structure so usually a layered structure is possible with relatively few regions where the ordering of the LC is frustrated [6]. If the boundary conditions for the director are perpendicular, the ordering is more frustrated. Extended line-like disordered regions are predicted close to the surface of the volume where the layered structure of the bulk of the droplet does not match the perpendicular boundary condition on the surface [7]. If the ordering has to emerge form a completely random disordered state, the layered chiral structure can trap some of the parts of the line-like disordered regions and prevent them from being expelled to the surface. This leads to three-dimensional arrangements of the line-like disordered regions which can become tangled and form non-trivial links and knots.

These predictions of complex structures in confined chiral nematic liquid crystals served as motivation for our study. In this Thesis we study director structures which can appear when chiral nematic LCs are confined to spherical volumes with perpendicular boundary conditions for the director field. Because of the complexity of such structures, the first step in the study was to develop a method capable of reconstructing director fields from experimental data. The basis of the new method was the well-established fluorescence confocal polarised microscopy [4, 8–10], which is very useful for examining director fields in 3D by projecting them to a direction perpendicular to the optical axis of the microscope. To achieve full characterisation of director fields, we augment the method by numerical optimisation [11] to find the missing sign of one of the components of the director field.

The Thesis is structured as follows: First we make a short introduction to liquid crystals and their topology. We then make a review of the literature on liquid crystal droplets with emphasis on chiral nematic droplets with perpendicular boundary conditions. This is followed by Chap. 4, where we introduce the microscopy methods that are used for examination of director fields in LC. Fluorescence confocal polarised microscopy is introduced and we discuss which information can be extracted from such measurements. In Chap. 5 we present the materials we use in our experiments, the sample preparation procedure and the experimental setup. Next, we develop a simulated annealing procedure in Chap. 6 and demonstrate its efficiency at reconstructing director fields of model chiral nematic droplets. We discuss the implementation of the simulated annealing algorithm on experimental data and demonstrate the procedure on several examples. In Chap. 7 we present some of the structures in chiral nematic droplets with homeotropic anchoring conditions. We explain how the string-like constellations of point defects and topological molecules can be constructed from simpler building blocks in Chap. 8. In Chap. 9 we review all the presented structures, discuss their metastability and compare the results with the predictions of the numerical studies. In the Conclusion, the main results are summarized and possible extensions and applications are discussed.

References

1. P.G. de Gennes, J. Prost, *The Physics of Liquid Crystals*, 2 edn. (Clarendon Press, Oxford, 1993)
2. I.I. Smalyukh, Y. Lansac, N.A. Clark, R.P. Trivedi, Three-dimensional structure and multistable optical switching of triple-twisted particle-like excitations in anisotropic fluids. Nat. Mater. **9**, 139–145 (2010)
3. P.J. Ackerman, Z. Qi, I.I. Smalyukh, Optical generation of crystalline, quasicrystalline, and arbitrary arrays of torons in confined cholesteric liquid crystals for patterning of optical vortices in laser beams. Phys. Rev. E **86**, 021703 (2012)
4. B.G.-G. Chen, P.J. Ackerman, G.P. Alexander, R.D. Kamien, I.I. Smalyukh, Generating the hopf fibration experimentally in nematic liquid crystals. Phys. Rev. Lett. **110**, 237801 (2013)
5. P.J. Ackerman, I.I. Smalyukh, Diversity of knot solitons in liquid crystals manifested by linking of preimages in torons and hopfions. Phys. Rev. X **7**, 011006 (2017)
6. D. Seč, T. Porenta, M. Ravnik, S. Žumer, Geometrical frustration of chiral ordering in cholesteric droplets. Soft Matter **8**, 11982–11988 (2012)
7. D. Seč, S. Čopar, S. Žumer, Topological zoo of free-standing knots in confined chiral nematic fluids. Nat. Commun. **5**, 3057 (2014)
8. I.I. Smalyukh, S. Shiyanovskii, O. Lavrentovich, Three-dimensional imaging of orientational order by fluorescence confocal polarizing microscopy. Chem. Phys. Lett. **336**, 88–96 (2001)
9. S. Shiyanovskii, I. Smalyukh, O. Lavrentovich, Computer simulations and fluorescence confocal polarizing microscopy of structures in cholesteric liquid crystals, in *Defects in Liquid Crystals: Computer Simulations, Theory and Experiments* (Springer, 2001), pp. 229–270
10. I. Smalyukh, O. Lavrentovich, Three-dimensional director structures of defects in Grandjean-Cano wedges of cholesteric liquid crystals studied by fluorescence confocal polarizing microscopy. Phys. Rev. E **66**, 051703 (2002)
11. S. Kirkpatrick, C. Gelatt Jr., M. Vecchi, Optimization by simulated annealing. Science **220**, 671–680 (1983)

Chapter 2
Liquid Crystals—An Overview

Liquid crystals are a well studied material, often serving as a model system for observing numerous physical phenomena. Here we will present only some of their basic properties and stress the subjects which will be useful in the interpretation of results and the discussion later in the Thesis. Most of the text here follows the classical works on liquid crystals such as Refs. [1, 2].

Liquid crystals is a name for a group of phases which flow under stress in a manner similar to liquids, but also possess order—either orientational or some degree of positional—like one would expect of crystalline solids. The building blocks of materials which exhibit LC phases are anisotropically shaped (most commonly rod- or disk-like), which causes ordering and results in anisotropic physical properties but the interactions between the building blocks are weak enough for the material to flow. The parameter which controls the phase of a liquid crystal is either temperature (thermotropic LC) or concentration (lyotropic LC).

There exist many phases of liquid crystals with varying degrees of orientational and positional order. The simplest phase is the nematic, which has only orientational order of its building blocks, with their centres of gravity being completely disordered, just as in an isotropic liquid. Two examples of a nematic phase are shown for rod-like (Fig. 2.1a) and disk-like (Fig. 2.1b) building blocks. An example of a phase with both positional and orientational order is the smectic A phase (Fig. 2.1c). In this phase the building blocks have an average direction similarly as in the nematic phase, but their density is modulated along one direction, which results in a layer-like structure. In each of the layers the building blocks are positionally disordered, forming a 2D isotropic liquid.

2.1 Nematic Order

In this Thesis we will focus on thermotropic liquid crystals with rod-like or calamitic organic molecules as their building blocks. In the simplest, nematic phase of liquid crystals the building blocks have an average orientation because of their shape but

© Springer Nature Switzerland AG 2018
G. Posnjak, *Topological Formations in Chiral Nematic Droplets*,
Springer Theses, https://doi.org/10.1007/978-3-319-98261-8_2

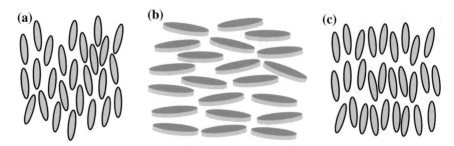

Fig. 2.1 Three examples of liquid-crystalline mesophases. **a** A nematic phase of rod-like or calamitic molecules. **b** A nematic phase of discotic molecules. **c** Smectic A phase with orientational order and positional order in one dimension

their centres of gravity possess no positional order, similar to a simple isotropic liquid. The most common way to describe their behaviour is to average out the orientation of the molecules on the microscopic scale to obtain a continuum description. This averaging yields a symmetric, traceless tensor Q, which can be written in the uniaxial approximation as the quadrupole moment of the distribution of molecular orientations [1]:

$$Q_{ij} = \frac{S}{2}(3n_i \, n_j - \delta_{ij}) . \tag{2.1}$$

The scalar *order parameter* S is the largest eigenvalue of Q and describes the degree of ordering in the system. It can be calculated as

$$S = \frac{1}{2}\langle 3\cos^2\theta - 1 \rangle , \tag{2.2}$$

where θ is the deviation of a molecule from the average direction and the chevrons denote averaging over the molecules. The eigenvector of Q which belongs to S is a normalised vector **n** called the *director* and corresponds to the average orientation of the molecules. Because Q in Eq. (2.1) is of second order in **n**, the director is a headless vector with the property **n** $=-$**n**, which corresponds to the building blocks of the LC having a symmetric head and tail [2].

Possible values of the order parameter S fall in the range $[-0.5, 1]$. In small molecule rod-like nematics they are typically between 0.3 and 0.7 [3–5]. In the disordered, isotropic phase the order parameter is equal to 0 and the director has no physical meaning. In a perfectly ordered, crystalline phase, the order parameter would be equal to 1 and all molecules would be aligned along the director, but still be positionally disordered.

2.2 Chiral Nematics

Closely related to the nematic phase is the cholesteric or chiral nematic phase (N*) in which the inversion symmetry is broken by chiral molecules (Fig. 2.2). Locally the

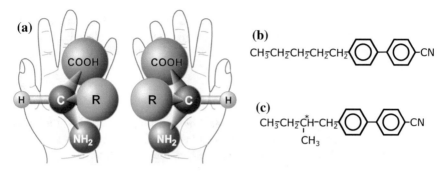

Fig. 2.2 Chirality in chemistry. **a** A molecule is chiral if its mirror image cannot be superimposed over it, similarly as left and right hands are mirror images of each other. For an organic molecule to be chiral, it is sufficient for it to have a carbon atom with four different atoms or groups of atoms connected to it. If a mesogenic molecule has no chiral atoms, as for example a 5CB molecule shown in panel (**b**), it forms achiral mesophases. A chiral version of 5CB molecule with the chiral C atom marked with an asterisk is shown in panel (**c**). Because of the chiral centre, the molecule has two enantiomers, the right-handed one called CB15 and the left-handed one, C15. Panel a is in public domain, reproduced from http://www.nai.arc.nasa.gov/

chiral nematic phase is similar to a normal nematic with the molecules on average pointing along a director and having no positional order of their centres of gravity. On a larger scale, however, the non-centrosymmetric molecules cause the director to be non-uniform in the N* ground state and slowly rotate along a direction perpendicular to it (Fig. 2.3). The distance over which the director completes a full rotation is called the pitch, p, and can vary over a large range from $0.1\,\mu$m upwards.

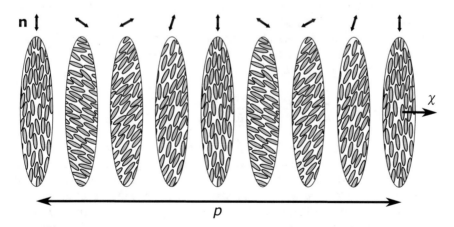

Fig. 2.3 Equilibrium structure of a bulk chiral nematic. Locally the molecules have an average direction and therefore a director can be determined. On a longer scale, the chirality of the phase induces rotation of the director field around an axis χ perpendicular to the director. The distance along χ over which the director rotates by 2π is called the pitch, p

Chirality can be an intrinsic property of the LC molecules as for example in cholesteryl ester (after which the phase was named), or it can be induced in a normal nematic by doping it with chiral molecules. In this case the pitch of the mixture is calculated as

$$p = \frac{1}{c \cdot HTP} \qquad (2.3)$$

where c is the concentration of the chiral dopant by weight and HTP is the helical twisting power of the dopant. HTP depends both on the properties of the dopant and the doped liquid crystal, and is usually temperature dependent.

2.3 Free Energy Expansion

In thermotropic LCs the order parameter depends on the temperature of the system. In the isotropic phase it is equal to 0, but it jumps discontinuously to a finite value in a first order phase transition when the temperature is lowered below the temperature of isotropic-to-nematic phase transition. This behaviour can be described by a phenomenological fourth-order Landau-de Gennes expansion of the free energy in terms of the Q tensor [2, 6, 7]:

$$f_0 = \frac{1}{2} A \operatorname{Tr} Q^2 + \frac{1}{3} B \operatorname{Tr} Q^3 + \frac{1}{4} C \left(\operatorname{Tr} Q^2 \right)^2 , \qquad (2.4)$$

where A is temperature dependent, and B and C are taken to be constant with $B < 0$. At high temperatures, where A is positive, the only minimum of f_0 is at $S = 0$. At lower temperatures, where A becomes negative, the equilibrium state moves to $S > 0$, corresponding to an ordered nematic phase.

The Landau-de Gennes expansion [Eq. (2.4)] gives only the free energy due to the ordering of the LC but does not include the energy of deformations of the director. These are included in the elastic free energy [7, 8]:

$$f_e = \frac{L_1}{2} \frac{\partial Q_{ij}}{\partial x_k} \frac{\partial Q_{ij}}{\partial x_k} + \frac{L_2}{2} \frac{\partial Q_{ij}}{\partial x_j} \frac{\partial Q_{ik}}{\partial x_k} + \frac{L_3}{2} Q_{ij} \frac{\partial Q_{kl}}{\partial x_i} \frac{\partial Q_{kl}}{\partial x_j} , \qquad (2.5)$$

where L_1, L_2 and L_3 are the tensorial elastic constants, which are properties of the material. An alternative way to write the elastic energy of a nematic LC at a constant temperature is to insert Eq. (2.1) into the equation above and consider S to be constant throughout the sample. This assumption is valid in most of the sample, where S changes only slightly, but breaks down in regions of suppressed order. Equation (2.1) can be then rewritten in terms of spatial variation of the director thus obtaining the Frank-Oseen elastic free energy [2]:

$$f_e = \frac{K_1}{2} (\nabla \cdot \mathbf{n})^2 + \frac{K_2}{2} (\mathbf{n} \cdot \nabla \times \mathbf{n})^2 + \frac{K_3}{2} (\mathbf{n} \times \nabla \times \mathbf{n})^2 . \qquad (2.6)$$

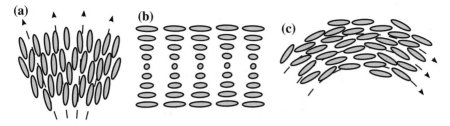

Fig. 2.4 The three modes of elastic deformation: **a** splay, **b** twist, and **c** bend

The three terms in this expansion correspond to the three basic modes of deformation of director structures: splay, twist and bend, all schematically shown in Fig. 2.4. K_1, K_2 and K_3 are their respective elastic constants which penalise each of the modes of deformation. The values of these constants depend on the material; typical values for small-molecule calamitic mesogens are on the order of 10 pN, with the twist constant being the smallest and bend usually the largest [5, 9, 10].

The elastic constants K_1, K_2 and K_3 are in this expansion of the free energy connected to the tensorial elastic constants by the relations [11]:

$$K_1 = \frac{9S^2}{4}(2L_1 + L_2 - L_3S) , \tag{2.7}$$

$$K_2 = \frac{9S^2}{4}(2L_1 - L_3S) , \tag{2.8}$$

$$K_3 = \frac{9S^2}{4}(2L_1 + L_2 + 2L_3S) . \tag{2.9}$$

Additional terms such as saddle-splay (K_{24}) and splay-bend (K_{13}) can be added to the expansion in Eq. (2.6), but they reduce to surface integrals which run over the boundaries of the nematic volume, i.e. the sample boundaries and the immediate vicinity of the defects. With strong enough homeotropic anchoring, the effect of these terms on the elastic behaviour near the outer surface of the LC volume becomes negligible. As we will deal mostly with experimental measurements, details of the elasticity around defects will play a smaller role and a single elastic constant approximation will suffice.

A common approximation in numerical modelling of LCs is to set the elastic constants to $K_1 = K_2 = K_3 = K$ with which we can derive the one-constant approximation of the elastic free energy [7]:

$$f_e = \frac{K}{2}[(\nabla \cdot \mathbf{n})^2 + (\nabla \times \mathbf{n})^2] . \tag{2.10}$$

The equivalent of this is setting the constants L_2 and L_3 to zero in the tensorial expansion of the free energy [Eq. (2.5)], which then becomes [6, 7]:

$$f_e = \frac{L_1}{2} \frac{\partial Q_{ij}}{\partial x_k} \frac{\partial Q_{ij}}{\partial x_k} . \tag{2.11}$$

In the case of chiral nematics the twist term in Eq. (2.6) has to be modified to reproduce spontaneous twisting in the equilibrium state:

$$f_{e,\text{TW}} = \frac{K_2}{2} (\mathbf{n} \cdot \nabla \times \mathbf{n} - q_0)^2 , \tag{2.12}$$

where q_0 is the inverse cholesteric pitch, $q_0 = 2\pi/p$. The director $\mathbf{n}$ has to twist in the ground state for this term to be equal to zero. In the expansion in terms of Q an additional term has to be added to introduce chirality:

$$f_{e,\text{TW}} = 2 q_0 L_1 \epsilon_{ikl} Q_{ij} \frac{\partial Q_{lj}}{\partial x_k} . \tag{2.13}$$

2.4 Anchoring

An important factor for the control of LC structures are the boundary conditions of the director field—the alignment of LC molecules on the edge of the LC volume. Because of elastic interactions, the orientation of the LC molecules on the surface can propagate deep into the bulk and dictate the structure and properties of LC. The orientation of molecules on the interface with another medium depends on the details of molecular interactions, but generally LCs with long alkyl chains are hydrophobic and their tails do not mingle with polar molecules such as water or glycerol. Interfaces with these two media have tangential or planar orientation of oil-like LC molecules (Fig. 2.5a). If there is no preferred direction on such an interface, the anchoring is called planar degenerate. On solid surfaces it is possible to brake this degeneracy. This can be for example done on polymer layers by rubbing them with a velvet cloth. The rubbing induces shallow grooves and partially orients the polymer chains on the surface, which both help to orient LC molecules and dictate a preferred direction [12].

Alternatively, on surfaces with long perpendicular apolar chains, the tails of LC molecules can penetrate such a layer and cause the LC director to be perpendicular

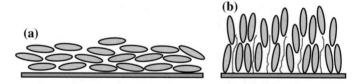

Fig. 2.5 Examples of orientation of LC molecules on interfaces: **a** tangential orientation is also called planar anchoring and **b** orientation normal to the interface is called homeotropic anchoring. The red chains in **b** are the long, non-polar tails of molecules which are attached to the substrate in order to induce homeotropic orientation of the LC molecules

to the interface. Such orientation is called homeotropic (Fig. 2.5b). There are also intermediate, tilted configurations of director. The tilt at the interface is generally difficult to measure, but it can significantly affect the topological properties of the director field [2, 13].

Anchoring of LC molecules can be modelled with a surface energy term, which penalises deviations of director **n** from a preferred direction $\mathbf{n}_0$, also called the easy axis. This therm is given by the Rapini-Papoular model [14, 15]:

$$f_{\text{anch}} = -\frac{1}{2} W (\mathbf{n} \cdot \mathbf{n}_0)^2 , \tag{2.14}$$

where the constant W determines the strength of anchoring. A characteristic anchoring length $l_{\text{W}} = W/K$ can be determined based on the energy of the anchoring and the elastic constants of the LC and is typically of the order of $1\,\mu$m [16]. In reality the deviations of orientation from the easy axis in the plane of the surface are less energetically costly than deviations in the direction perpendicular to the surface. The measurements of anchoring energy give values between 10^{-7} and 10^{-3} Jm^{-2} for W_{polar} and $W_{\text{azimuthal}}$ smaller by one or two orders of magnitude [12].

The anchoring of LC molecules at an interface can be modified by adding a surfactant. For example, amphiphilic molecules such as SDS and CTAB have a polar part which is hydrophilic and long non-polar tails which are hydrophobic. The polar part orients to the water side of the interface and the long, non-polar tails penetrate into the LC volume and orient the LC molecules (Fig. 2.5b). By adding such surfactants the type and strength of anchoring can be tuned [2, 17, 18].

2.5 Topological Defects in Nematics

As we have shown, elastic energy of a liquid crystal depends on the deformations or gradients in the director field. Let us imagine a liquid crystal confined to a spherical volume, with molecules on the surface of the sphere being oriented perpendicularly to the interface (homeotropic anchoring; Fig. 2.6a). If we extrapolate the orientation of the director toward the centre of the volume, we find that the splay term in the elastic energy will become increasingly large. This term can become so large that the elastic energy can locally increase to a point where it is energetically favourable for the phase to change to isotropic and thus avoid the high energetic cost of the elastic deformation. In the centre of a LC confined to a spherical volume with homeotropic anchoring we will therefore find a small isotropic island where the director is not defined—a defect in the director field with a size on the order of 10 nm [19]. For the same reason a defect would appear in a cube-like volume with homeotropic anchoring on all surfaces but would be absent if only two opposite surfaces were homeotropic and the rest were planar (Fig. 2.6b). We can see that the presence of such a defect does not depend on the geometrical shape of the confinement but on something more general—its topology. Therefore such defects are called topological defects.

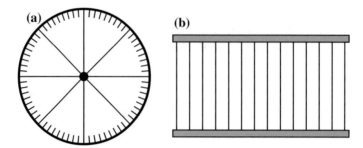

Fig. 2.6 Examples of director fields in confined volumes. **a** If we extrapolate the director in a spherical volume of LC from the homeotropic alignment on the surface, we find that the director is undefined in the centre of the volume, where because of the high free energy cost of the elastic deformation the medium locally melts to isotropic phase. **b** In a volume with homeotropic anchoring on two opposite parallel sides and planar or no anchoring on the other sides no such defect is present

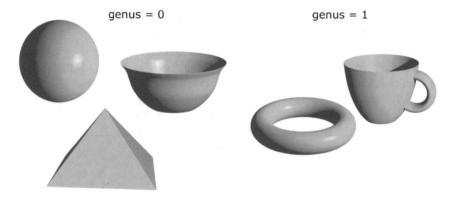

Fig. 2.7 Examples of solid bodies with different topologies. A sphere, a pyramid and a bowl are all topologically equivalent because they can be smoothly deformed into each other without cuts. For example, a sphere can be flattened into a disc and then bent into a bowl-like shape. Because these solids have no holes, their genus is equal to 0. On the other hand, both a torus and a mug have a single hole which punctures through their volume and therefore their genus is equal to 1

Topology studies the connectedness of surfaces and volumes, and all objects that can be smoothly deformed into each other without cuts are considered topologically equivalent. Because holes cannot be introduced or removed without cuts, bodies are topologically classified with regard to the number of holes they have. From a topological point of view a sphere is therefore equivalent to a cube, a pyramid, or a bowl, but not to a mug with a handle, because the handle introduces a hole which punctures the volume of the body (Fig. 2.7). The number of holes in an object is called genus (g) and for closed surfaces it is connected to the Euler characteristic of the object $\chi = 2(1 - g)$, which can be calculated from the total curvature of the object through the Gauss-Bonnet theorem [20]:

$$2\pi \chi = \oint \kappa \, \mathrm{d}S = \oiint \mathbf{v} \cdot (\partial_\vartheta \mathbf{v} \times \partial_\varphi \mathbf{v}) \, \mathrm{d}\varphi \, \mathrm{d}\vartheta \, , \tag{2.15}$$

where κ is the local Gaussian curvature of an infinitesimal patch of surface $\mathrm{d}S$, $\mathbf{v}$ is the normal of the surface, and φ and ϑ are parameters which run over the entire surface. We will use these relations later to determine topological constants of director fields.

2.5.1 Topological Defects in Two Dimensions

Topological defects in director fields are classified by the properties of the director field which surrounds them. If a nematic is confined to a 2D surface (the director lies completely in one plane), a defect is classified by the number of rotations the orientation of the director performs when the defect is encircled once along an arbitrary closed path—its winding number, defined as [2]:

$$k = \frac{1}{2\pi} \oint_\gamma \frac{\mathrm{d}\varphi}{\mathrm{d}l} \mathrm{d}l \, , \tag{2.16}$$

where γ is the closed path and φ is the orientation of the director field in the plane. A few examples of defects with different winding numbers are shown in Fig. 2.8. Defects with half-integer winding numbers where the director rotates by π are allowed in 2D because of the nematic symmetry $\mathbf{n} = -\mathbf{n}$.

An interesting result which can be calculated from Eq. (2.10) for 2D defects is that the free energy of the elastic deformation they induce is proportional to the square of their winding number which has important implications for their stability [1].

Two-dimensional defects can also appear on surfaces of otherwise 3D samples. Such defects are common in situations with degenerate planar anchoring [13]. If they have a half-integer winding number, they are terminal points of disclination lines, and if they have integer winding numbers, they are point defects called boojums. An example of such a defect is shown in Fig. 2.9a and an example of a 3D director configuration in a LC droplet with degenerate planar anchoring is shown in Fig. 2.9b.

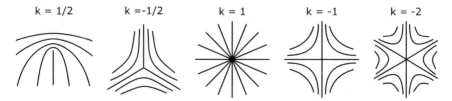

Fig. 2.8 Two-dimensional topological defects with different winding numbers. A negative winding number means that the director rotates in the opposite sense than the path encircling the defect

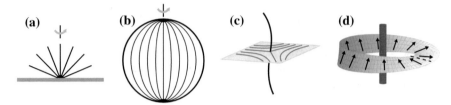

Fig. 2.9 Defects in 3D director fields. **a** A surface defect with $k = 1$, called a boojum. **b** An example of a nematic droplet with planar degenerate anchoring. A bipolar structure satisfies the anchoring almost everywhere on the surface except in two points where boojums are formed. **c** An example of a disclination line. The director field in a plane perpendicular to the disclination line lies in-plane, is therefore 2D and has $k = -1/2$ winding number. **d** An example of a twist disclination. If we follow the rotation of the director along a closed path around the disclination (starting at the dashed line), the director first rotates out of the cross-section plane and then continues the rotation until it is again lying in-plane. During this rotation it traces a Möbius strip as indicated in the image. From the orientation of the arrows we prescribed to the director, we can see that it rotates by π

The sum of the winding numbers of defects of a director field with non-zero tangential components on a closed surface is a conserved quantity: the Poincaré-Hopf theorem states that it is equal to the Euler characteristic of the surface, which can be calculated from Eq. (2.15) [16, 21]. This means that on a spherical particle with planar anchoring the defects must have a total winding number equal to 2 and on a torus-like particle with planar anchoring, the sum of the winding numbers of defects must be 0.

The winding number can also be used to classify line defects—extended tube-like isotropic regions. If the director field lies in a plane, perpendicular to the disclination, the line disclination is analogous to a 2D defect and can be classified by the winding number of its director profile. An example of a disclination line with $k = -1/2$ is shown in Fig. 2.9c. Because line defects are embedded in 3D, the director can also rotate out of the plane of their cross-section. In this case their topological classification can be done by a generalized form of the winding number: instead of observing the rotation of the director in the plane of the cross-section, we imagine the director is drawn on a ribbon and we can see the ribbon twists by π while it encircles the defect core once (Fig. 2.9d). In nematics in general there is no restriction for the director in the cross-section of a disclination with half-integer winding to be in-plane and the winding profile of a disclinations can change. Because of this, all line defects with a half-integer winding number director profile, which does not change along the length of the disclination, are topologically equivalent [22–24].

2.5.2 Three-Dimensional Topological Defects

Singular points in the bulk of 3D director fields (point defects) are characterised by their topological charge, which is defined analogously to the winding number [22]:

$$q = \frac{1}{4\pi} \oiint \mathbf{n} \cdot \left(\partial_\vartheta \mathbf{n} \times \partial_\varphi \mathbf{n} \right) \mathrm{d}\vartheta \, \mathrm{d}\varphi \,, \tag{2.17}$$

where ϑ and φ are a parametrisation of a closed surface encompassing the point defect, and the integral runs over the entire surface. To calculate q, we have to treat $\mathbf{n}$ as a vector—we have to assign an orientation to it, or in other words, add an arrow to it. Equation (2.17) is a mapping of the director field $\mathbf{n}$ to a unit sphere—it counts how many times the director on a surface enclosing the point defect visits each possible orientation on a unit sphere. Because q depends on the third power of $\mathbf{n}$, the sign of topological charge depends on the direction we assign to the director. By transforming an otherwise headless director to an arrow, we are breaking the symmetry of the nematic. In this way we produce negative and positive topological defects. However, since the choice of direction was arbitrary, the sign of the topological charge is arbitrary as well. In practice this means that if we want precise bookkeeping of the signs of topological charge in a sample, e.g., so that we can calculate the total topological charge, we need to have a common reference. Once we add an arrow to the director at some point in volume, we thus have to take care that this choice is consistent in all the volume.

If we take care of consistent arrow assignment throughout the LC volume, the total topological charge of all the defects can be calculated with simple addition. In the case of homeotropic anchoring on the surface of the LC volume, the director $\mathbf{n}$ is equal to the normal to the surface $\boldsymbol{\nu}$. Therefore Eq. (2.17) becomes almost identical to Eq. (2.15) and we can extract a simple relation for the total topological charge inside the volume: $q = 1 - g$. Total topological charge is a conserved topological quantity and depends solely on the genus g or in other words, the number of holes in the LC volume. This expression also works the other way—if we place a particle with a genus g and homeotropic anchoring on its surface inside a LC volume, the director around it will behave as if a defect with the topological charge q is induced.

A quick way of identifying the sign of charge we have chosen for unit charge defects is to observe the direction of arrows on the symmetry axis of the defect. If the arrows are pointing away from the defect as in Fig. 2.10a, its charge is positive, and if they are pointing towards the defect (Fig. 2.10b), it is negative—regardless of the geometrical shape of the surrounding director. The familiar source-and-sink-like behaviour of topological charges could lead us to drawn parallels between the director field and its topological charges and the electric field lines and electrostatic charges. However, the analogy is not justified because of fundamental differences between the two: the director field is, unlike electric field, headless and normalised, and topological charge is not calculated as the divergence of the director field lines as in the electrostatic case, but as the mapping of a director field to a unit sphere. Hyperbolic point defects are consequently not proper sinks of the director field but more akin to saddle points in the electric field which appear between electric charges of equal sign [25].

Equation (2.17) is in essence just a calculation of how many times the director on a surface enclosing the defect covers all possible orientations in space. This can be quickly deduced by decomposing the director on the surface surrounding the

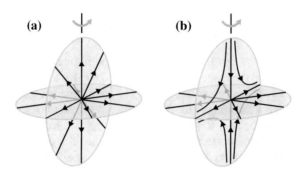

Fig. 2.10 Point defects in 3D director fields. **a** A radial and **b** a hyperbolic point defect. Both point defects are rotationally symmetric around the vertical axis. In the equatorial plane the director field of both defects is radial, but in all the vertical planes the director field of the defect in **b** has a hyperbolic configuration. Arrows are added to help determine the topological charge of the defects

defect into patches by projecting it to the normal of the surface. With this, patches of "in-going" and "out-going" director are formed, separated by boundaries where the director rotates by π. These boundaries can include "grains" where the director changes direction by an in-plane rotation. The topological charge of a defect can thus be calculated as [26]:

$$q = 1 + \sum_{i}^{N} t_i + \sum_{i}^{M} g_i \qquad (2.18)$$

where t_i of the N patches can be ± 1, depending on the orientation of the director on the patch and g_i of the M grains takes values of $\pm 1/2$, depending on the direction of rotation of the director in the grain.

Unlike 2D defects for which winding numbers larger than one have been observed experimentally under certain conditions [27–29], 3D point defects in nematics have so far only been observed in two distinct flavours: radial (with its twisted variants) and hyperbolic, both possessing unit topological charge [30]. If a higher defect would appear in a nematic, theoretical predictions show it would not be stable [31] and experiments which tried to induce higher-charge defects observed they indeed dissociate into a suitable number of unit defects [32].

Within a topological genus, different defects can appear. For example, in a spherical droplet with homeotropic anchoring, the point defect we showed in Fig. 2.6a can be replaced by a closed line defect—a defect loop. These two defects are equivalent if they have the same topological charge. Topological restrictions do not dictate the exact type or number of defects, only their total topological charge. Stability of different defects is then determined by their free energy. Differences between free energies of different defects arise because of contributions of the elastic energy of the deformed director field surrounding the defect and the energy of the defect core, which is proportional to its volume and is larger for line defects than point defects [30, 33].

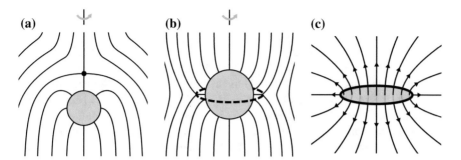

Fig. 2.11 Spherical particle in a uniform director field. If a spherical particle with homeotropic anchoring is put in a uniform director field, it will carry a topological charge of +1 because of the radial orientation of the director around it. Its topological charge needs to be compensated by another, induced, defect that can be either **a** a point defect or **b** a ring defect. Both induced defects carry a unit topological charge of the opposite sign, so that the total topological charge is 0. **c** A loop defect with a half-integer winding number of its director profile induces a discontinuity in the prescription of arrows to the director. This can be resolved by imagining a membrane (branch cut surface) stretched across the ring (shaded area) on which the arrows change direction. In the shown case, the arrows on the symmetry axis of the membrane are pointing away from the membrane, and the disclination has a +1 topological charge

These energetic differences dictate which defects will appear in a given situation. For example, if we put a solid spherical particle with homeotropic anchoring on its surface into LC with uniform director field, a companion defect is induced to compensate the topological charge of the particle [34]. If we choose the arrows on the surface of the particle to point away from the particle, its topological charge will be +1. The companion defect therefore needs a topological charge of −1 to compensate the topological charge of the particle, as their combined charge needs to be equal to 0 because the far-field director is uniform. This defect can be realized in two ways—either with a hyperbolic point defect (Fig. 2.11a) or a loop with a half-integer winding number (Fig. 2.11b). Which defect will appear is strongly influenced by the confinement of the system. If the distance between the bounding surfaces of the LC cell is comparable to the size of the particle, a defect ring around the particle will be the stable configuration. On the other hand, if the distance is increased, the energy of the defect core will be minimised by shrinking the loop into a point defect [35].

Problems with assigning an arrow to an otherwise headless director arise if the LC volume contains line defects with half-integer winding numbers, which are allowed in nematics. If an arrow is added to the director in such cases, it flips direction after encircling the defect core once and we therefore cannot find an assignment of arrows which would be consistent throughout the volume [22]. We can rectify this situation by imagining a membrane or a branch cut surface [36], stretched over the opening in the defect loop as shown in Fig. 2.11c. If we suppose the vectorised director flips its orientation when it passes through the membrane, the discontinuity in the oriented director field is resolved. By doing this we can also determine the sign of topological charge of the simple defect loop in Fig. 2.11c in the same way we did with point defects. A disclination line with non-rotating half-integer winding number of its cross-section carries a +1 topological charge if the arrows at the membrane are pointing outward as in Fig. 2.10c, and a −1 topological charge if they point inward. If the LC volume includes point defects in addition to the line defects, the situation is changed; namely, the assignment of the signs of topological charges of the point defects depends on the choice of the branch cut surface [22]. Because the total topological charge is determined by the topology of the confinement, and the anchoring conditions, the choice of the branch cut surface also changes the topological charge of the ring defect. Without selecting a branch cut surface, one can still tell if the topological charge of a defect line is odd or even: if the director profile of the defect line with half-integer winding is fixed, the topological charge is odd, and if the profile changes, it can be odd or even, depending on the number of changes [23, 37]. Without a selected branch cut surface to define the signs of the topological charges, the conserved topological quantity is the sum of the topological charges modulo 2 [22, 23, 38, 39].

The topological charges of the spherical particle and the companion defect are elastically attracted to each other. This attraction arises because areas of deformed director around defects carry elastic energy and this energy can be lowered when two deformed areas overlap [33]. If the deformation in the two areas is not geometrically compatible, the elastic energy can increase when the two defects approach each other, resulting in a repulsive elastic force. Repulsion usually arises between defects of the same type [24]. The elastic force can have a range of several tens of micrometers [40].

Two defects with opposite topological charge attract if they are placed into a homogeneous director field [33, 34, 40–43]. They will slowly approach each other and annihilate when they collide. Such situations arise for example when a LC sample is quenched from the isotropic to nematic phase [44–46]. In the first moments after the phase transition the sample is in the nematic state, but the information about the alignment has not propagated through the sample yet. Small domains of uniform director are formed, and as they merge, a dense network of defects is formed, where the orientations of the merging domains do not match. Most of these defects annihilate quickly, leaving behind only the defects which are necessary because of topological constraints of the confinement. For example, in a volume with uniform director alignment on its surface (Fig. 2.6b), a pair of oppositely charged unit topological defects will annihilate into the uniform director state just as a single defect ring will, because it needs to be topologically trivial due to the boundary conditions. In a droplet with homeotropic anchoring the situation is different—the topology of confinement dictates that the total charge inside the volume must be equal to 1 and therefore at least one point or loop defect will be stable, as seen in Fig. 2.6a. A way to stabilize defects after a quench in uniform field is to add inclusions which prevent their annihilation—particles of any shape [24, 47–51], for example as seen in Fig. 2.11a, b.

In some cases the defects which satisfy topological constraints are metastable—their energy is higher than that of the most stable configuration, yet the state is stable because the energy barrier between the states is significantly higher than the energy of thermal fluctuations [37, 38]. In such cases, any of the metastable states

(a)   **(b)**   **(c)**

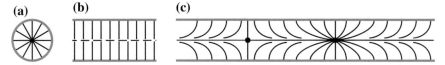

Fig. 2.12 Director field in cylindrical capillaries with homeotropic anchoring. A cross-section oriented **a** perpendicularly to and **b** along the axis of the capillary in which the director lies completely in the perpendicular cross-section in (**a**). Because of radial configuration of the director in this plane, a singular disclination line with $+1$ winding number is needed along the axis of capillary. **c** Situation in which the director is not singular along the axis, because it escapes by bending out of the plane of the perpendicular cross-section. Both directions of escape along the axis are possible and where the areas of differently oriented escape meet, hyperbolic or radial point defects are formed

can arise if the system is quenched from the isotropic phase. In some situations, the most symmetric structure features a defect, but if the symmetry is broken, a more stable structure without a defect is attainable. An example is the director field in a cylindrical capillary with homeotropic anchoring on the surface [30, 52]. The most symmetric structure in this system would be a radial director field in the cross-section of the capillary (Fig. 2.12a) with a $k = +1$ line defect in the centre extending along the symmetry axis of the capillary (Fig. 2.12b). However if the director tilts out of the plane of the cross-section in Fig. 2.12a, a structure without a singular defect line is possible. In this case the director is perpendicular on the surface of the capillary, but it avoids the singularity along the axis of the tube by bending out of the plane of the perpendicular cross-section as can be seen at the left end of the capillary in Fig. 2.12c. The singular core of the more symmetric structure is in this case substituted with non-singular director pointing along the axis, splaying and bending toward the surface of the capillary. Such a resolution of a singular structure is commonly called "escape along the third dimension". Which of the two possible structures is energetically favourable depends on the balance of absence of the extended defect core lowering the free energy and extra splay and bend deformation of the director field increasing it. The energy balance is dictated by the elastic constants of the LC and its confinement—anchoring strength and the diameter of the capillary. Because the director can escape in any of the two possible directions, domains with opposite directions of escape are separated by ± 1 point defects as seen in Fig. 2.12c. Depending on the dynamics of the quench many or only a few defects can form.

2.5.3 Topological Defects in Chiral Nematics

In a chiral nematic liquid crystal the variety of line defects is greater, but we will only briefly go through them. The special line defects appear only in proper cholesteric samples with a well-defined cholesteric axis almost everywhere in the sample. As we will see later, these types of defects are not present in the studied chiral nematic droplets because the confinement greatly frustrates the cholesteric structure.

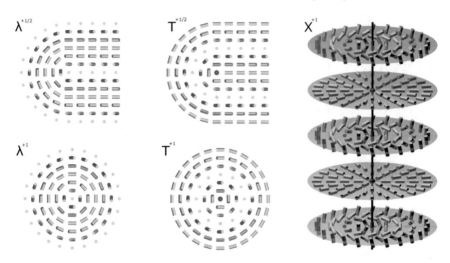

Fig. 2.13 Disclinations in cholesteric samples. In each type of disclination two out of the three directions which locally characterise the cholesteric are not defined and the type of disclination is named after the defined direction and the winding of the disclination. For the χ^{+1} disclination, the director in a series of cross-sections along the core of the disclination is shown

To properly characterize cholesteric structures two more directions besides the director need to be defined—the direction of the chiral axis around which the director rotates, χ, and a direction which is perpendicular to both the director and the chiral axis, τ [2]. Areas where any pair of these three directions is undefined form cholesteric line defects, examples of which are shown in Fig. 2.13. Disclinations in cholesterics are named after the vector which is not singular at their core, with disclinations with non-singular director being called λ disclinations. All of these cholesteric defect lines are properly defined only in bulk cholesteric LC where the surrounding volume is filled with cholesteric layers—they are discontinuities in the layers. At the edges of cholesteric volume, where a part of the volume is nematic, cholesteric defects do not appear or are not localized, making it difficult to define the defect core. For example, in a planar wedge cell filled with a cholesteric, the discontinuity between the thinnest region with zero twist and the first cholesteric region with π twist is not one of the cholesteric defects but a twist disclination.

Point defects only rarely appear in cholesteric samples with well-defined helical structure. One such example is the nucleation of a series of point defects from a χ disclination line which runs perpendicularly to the cholesteric layers. The singular director of the line escapes in the third dimension similarly as in a capillary, but the periodic helical structure of the cholesteric layers organises the alternating ± 1 defects into a string with regular spacing [1]. On the other hand, in systems where the confinement frustrates the helical ordering, the whole phase can be considered as the border between the nematic and cholesteric. For example, in a cholesteric LC with the pitch comparable to the distance between two surfaces with homeotropic anchoring, localised regions of twist called bubble domains or torons will be separated from the homeotropic surface by point defects [53–56].

2.6 Optics of Liquid Crystals

Liquid crystals strongly interact with light either through their anisotropic dielectric properties or through absorption and non-linear processes. This makes LC closely connected with optics—optical methods are one of the most widely used tools for observation of their structure and photonic applications are their most common uses. The basis of our observation of LC structures is microscopy so it is important to understand the optical properties of LCs.

To describe light as an electromagnetic wave, we need four fundamental quantities: electric field **E**, electric displacement **D**, magnetic field **H** and magnetic-flux density **B**, which are connected through the Maxwell equations. For light in a medium without free charges, the Maxwell equations can be written as [57, 58]:

$$\nabla \cdot \mathbf{D} = 0 , \tag{2.19}$$

$$\nabla \cdot \mathbf{B} = 0 , \tag{2.20}$$

$$\nabla \times \mathbf{E} = -\frac{\partial \mathbf{B}}{\partial t} , \tag{2.21}$$

$$\nabla \times \mathbf{H} = \frac{\partial \mathbf{D}}{\partial t} , \tag{2.22}$$

with additional relations between the fields in matter:

$$\mathbf{D} = \varepsilon_0 \, \underline{\varepsilon} \, \mathbf{E} , \tag{2.23}$$

$$\mathbf{B} = \mu_0 \, \underline{\mu} \, \mathbf{H} , \tag{2.24}$$

where $\varepsilon_0 = 8.85 \times 10^{-12}$ (As)^2/Nm^2 is the vacuum permittivity, $\underline{\varepsilon}$ the dielectric tensor of the medium, $\mu_0 = 4\pi \times 10^{-7}$ Vs/Am the vacuum permeability and $\underline{\mu}$ the permeability tensor of the medium. For small-molecule organic LCs, $\underline{\mu}$ is equal to 1 and they can be considered non-magnetic. For uniaxial, non-absorbing LC, the dielectric tensor in its principal frame can be written as

$$\underline{\varepsilon} = \begin{pmatrix} \varepsilon_1 & 0 & 0 \\ 0 & \varepsilon_1 & 0 \\ 0 & 0 & \varepsilon_3 \end{pmatrix} , \tag{2.25}$$

where ε_3 is the dielectric permittivity along the director of the LC, and ε_1 in a direction perpendicular to the director. In such a medium, the electric field **E** and magnetic field **H** are both perpendicular to the direction of light propagation given by the wave vector **k**, and to each other. Because of this, an electromagnetic wave can be fully characterised by its electric field: $\mathbf{E} = \mathbf{E}_0 \exp(\mathbf{k} \cdot \mathbf{r} - \omega t)$. The magnitude of the electric field $\mathbf{E}_0$ is in general a complex vector and defines the polarisation of light. If the polarisation is a real vector and points along one of the principal axes of the dielectric tensor, it does not change its orientation while propagating through the material; such a wave is an eigenmode of the material, which does not change polarisation. For such light, the wave vector can be written as $\mathbf{k} = 2\pi/\lambda_0 n_{e,o}$, where λ_0 is the wavelength of light in vacuum, $n_e = \sqrt{\varepsilon_3}$ is the refractive index of the material for a polarisation along the LC director and $n_o = \sqrt{\varepsilon_1}$ is the refractive index for any polarisation perpendicular to the director, where $\varepsilon_{1,3}$ are the relative dielectric constant values of the material at the frequencies of the propagating wave. Because ε changes with the wavelength of light, liquid crystals have dispersion.

Light propagating along the director sees the same ordinary refractive index, n_o, for all polarisations. If it is propagating at a perpendicular direction with regard to the director, its polarisation can be decomposed into two perpendicular components, one with ordinary refractive index and the other with extraordinary refractive index, n_e. Because of the different refractive indices for the two components of polarisations, light with different polarisations gathers a different phase difference, also called retardance $\delta = \mathbf{k} \cdot \mathbf{r}$, while propagating through the medium. Because of the difference in retardance, the material is birefringent and can change the polarisation of light, propagating through it.

If the vector $\mathbf{E}_0$ is real, the polarisation of light is linear. If it is complex, the two components oscillate with different phases and the beam has an elliptical polarisation. Propagation through a birefringent medium changes the relative phase of the two components and with it the polarisation state of light.

For light propagating at an oblique angle to the LC director, the polarisation of light is still decomposed in two components: one in a direction perpendicular to the director with refractive index n_o and the other one perpendicular to this direction and to the direction of light propagation, for which the effective index has to be calculated from the refractive index ellipsoid [58, 59]. In such a case, the refractive index of the extraordinary wave is calculated as [59]

$$\frac{1}{n^2} = \frac{\cos^2 \vartheta}{n_o^2} + \frac{\sin^2 \vartheta}{n_e^2} \, , \tag{2.26}$$

where ϑ is the angle between the wave vector $\mathbf{k}$ and the director.

There are several methods for calculating the influence of an optically inhomogeneous material on the polarisation state of a light wave, the simplest being the Jones matrix method [59], which can handle relative phase changes between two orthogonal polarisations, but not reflections. A more complex method is the Berreman 4×4 matrix method which can also handle reflections [60]. In both these methods you can write a matrix to describe a homogeneous optical material and this matrix then acts upon an appropriate vector which describes the polarisation state of light. If an optically inhomogeneous material is considered, it has to be divided into a series of thin slices, each described with its own matrix, which act upon the polarisation state vector in the same sequence as the light passes through the layers.

Another related method is the Mueller matrix method which acts upon the Stokes vectors [58]. The Stokes vector is a generalisation of Jones vector, with which we can also describe unpolarised light. With Stokes vectors, it is possible to determine the polarisation state of light from experimentally measured intensities. We will

Table 2.1 Examples of polarisation states

Polarisation	Vector notation	
	(E_x, E_y)	Stokes vector
Unpolarised	/	$(1, 0, 0, 0)$
Linear $(0°)$	$(1, 0)$	$(1, 1, 0, 0)$
Linear $(90°)$	$(0, 1)$	$(1, -1, 0, 0)$
Linear $(45°)$	$(\sqrt{2}/2, \sqrt{2}/2)$	$(1, 0, 1, 0)$
Linear $(135°)$	$(\sqrt{2}/2, -\sqrt{2}/2)$	$(1, 0, -1, 0)$
Circular (righthanded)	$(\sqrt{2}/2, i\sqrt{2}/2)$	$(1, 0, 0, 1)$
Circular (lefthanded)	$(\sqrt{2}/2, -i\sqrt{2}/2)$	$(1, 0, 0, -1)$
Elliptical (righthanded with long axis along x)	$(0.8, i0.6)$	$(1, 0.28, 0, 0.96)$
Elliptical (righthanded with long axis along y)	$(0.6, i0.8)$	$(1, -0.28, 0, 0.96)$

present this description in more detail, because it will be useful in a later chapter, when we will explain, how we calculate the orientation of director from fluorescence intensities, measured at different polarisations.

If the polarisation of the electric field is $\mathbf{E} = (E_x, E_y)$, the four Stokes vector components can be written as [58]

$$S_0 = \langle E_x E_x^* + E_y E_y^* \rangle \,, \tag{2.27}$$

$$S_1 = \langle E_x E_x^* - E_y E_y^* \rangle \,, \tag{2.28}$$

$$S_2 = \langle E_x E_y^* + E_y E_x^* \rangle \,, \tag{2.29}$$

$$S_3 = \langle i(E_x E_y^* - E_y E_x^*) \rangle \,, \tag{2.30}$$

where the brackets denote the value averaged over time. S_0 is the light intensity, S_1 the difference between the components in x and y direction, S_2 the difference in components along the $\pm 45°$ directions, and S_3 the difference in the right and left handed components of circular polarisation. If $(S_1)^2 + (S_2)^2 + (S_3)^2 < (S_0)^2$, the light is not completely polarised. Several examples of polarisation states are given in Table 2.1

2.6.1 Optics of Chiral Nematic Liquid Crystals

We already discussed how the director in chiral nematic LCs rotates around a helical axis. For example, the equilibrium configuration of a cholesteric in bulk, where the influences of the surfaces cannot be felt, can be written as [1]:

$$\mathbf{n} = \big(\cos(2\pi z/p), \sin(2\pi z/p), 0\big) \,, \tag{2.31}$$

where the chiral axis χ is aligned with the coordinate axis z and p is the pitch of the cholesteric.

Such a sample is not optically homogeneous, because the principal axes of the dielectric tensor follow the director. If light is travelling along the layers, the spatially varying refractive index effectively acts as a lens, locally focusing or dispersing the light. The deviation of a light beam from its original path because of this lensing can be estimated as $z\Delta n$ where Δn is the birefringence of the LC and z is the depth of scanning [61, 62].

If we want to consider propagation of a light wave along the chiral axis χ (the z axis in this case), we need to cut the sample into a series of xy slabs with a thickness much smaller that p, so that the director in a slab does not rotate considerably and the slab can be considered to have homogeneous director. Then we can model the propagation of light through the LC with a series of matrices corresponding to the xy slabs, each with its principal axis rotated slightly with respect to its neighbours.

For light propagating along the helical axis, the cholesteric described in Eq. (2.31) acts a polarisation state transformer. Several regimes of operation can be derived, depending on the refractive indices of the medium, its pitch and the wavelength of light. In each regime, two polarisation states of light can be found, which are eigenmodes of the cholesteric structure—they do not change while propagating through the medium. The eigenmodes are in general elliptically polarised, with the axes of the ellipses following the local principal axes of the dielectric tensor [59]. The four limit regimes, which depend on the wavelength of light λ, chiral pitch of the structure p, average refractive index n and the birefringence Δn, are [59]:

1. Mauguin regime ($\lambda \ll 0.5\, p\, \Delta n$). In the short wavelength spectral regime, where the director rotates slowly compared to the wavelength of light, the two eigenmodes are linearly polarised, with polarisation of one pointing along the director and of the other perpendicular to the director. Because the polarisation of both modes follows the local director, this regime can be used for polarisation guiding, for example in twisted nematic cells and other LC devices [58].
2. Short wavelength circular regime ($0.5\, p\, n\, \Delta n \ll \lambda \ll p$). In this spectral regime, the eigenmodes have nearly circular polarisation state. Circularly polarised light of either handedness does not change its polarisation state in this regime. Light with arbitrary polarisation state is decomposed into a linear combination of the left- and right-handed circular polarisations, but because the two eigenmodes propagate through the medium with a different wavevector $\mathbf{k}$, the polarisation state of incoming light changes. For a linear polarisation this means the plane of polarisation is rotated by rotatory power [59]:

$$\rho = -\frac{\pi(\Delta n)^2 p}{4\lambda^2}. \tag{2.32}$$

3. Bragg regime ($n_o\, p < \lambda < n_e\, p$). In the Bragg regime the pitch of the cholesteric matches the wavelength of light in the medium. The two eigenmodes are circularly polarised. The eigenmode with the opposite handedness as the cholesteric

structure sees a periodic structure, which acts as a Bragg mirror. Because of the matching of periodicity of the structure and the wavelength, the reflected waves interfere constructively. The wavevector of such a polarisation state is purely imaginary, so only an evanescent wave is present in the medium, and all light with a handedness opposite to the cholesteric is reflected. The polarisation state with the same handedness as the structure does not see the periodic structure of the cholesteric, so it can propagate through the medium.

4. Long wavelength circular regime ($n_e\, p \ll \lambda$). In the long wavelength circular regime both eigenmodes are almost circularly polarised. Because of the differences in the wavevectors of the two eigenmodes, the rotatory power in this regime is [59]:

$$\rho = \frac{\pi n^2 (\Delta n)^2 p^3}{4\lambda^4}. \tag{2.33}$$

References

1. P.G. de Gennes, J. Prost, *The Physics of Liquid Crystals*, 2nd edn. (Clarendon Press, Oxford, 1993)
2. M. Kléman, O.D. Lavrentovich, *Soft Matter Physics: An Introduction* (Springer Science & Business Media, 2003)
3. P. Karat, N. Madhusudana, Elastic and optical properties of some 4-n-alkyl-4-cyanobiphenyls. Mol. Cryst. Liq. Cryst. **36**, 51–64 (1976)
4. M.L. Magnuson, B. Fung, J. Bayle, On the temperature dependence of the order parameter of liquid crystals over a wide nematic range. Liq. Cryst. **19**, 823–832 (1995)
5. A. Srivastava, S. Singh, Elastic constants of nematic liquid crystals of uniaxial symmetry. J. Phys.: Condens. Matter **16**, 7169 (2004)
6. M. Ravnik, G.P. Alexander, J.M. Yeomans, S. Žumer, Mesoscopic modelling of colloids in chiral nematics. Farad. Discuss. **144**, 159–169 (2010)
7. S. Čopar, Topology and geometry of defects in confined nematics. Ph.D. thesis, Faculty of mathematics and physics, University of Ljubljana, Slovenia, 2012
8. L. Longa, D. Monselesan, H.-R. Trebin, An extension of the Landau-Ginzburg-de Gennes theory for liquid crystals. Liq. Cryst. **2**, 769–796 (1987)
9. P. Karat, N. Madhusudana, Elasticity and orientational order in some 4-n-alkyl-4-cyanobiphenyls: Part II. Mol. Cryst. Liq. Cryst. **40**, 239–245 (1977)
10. H. Schad, M. Osman, Elastic constants and molecular association of cyano-substituted nematic liquid crystals. J. Chem. Phys. **75**, 880–885 (1981)
11. D. Seč, Ordering and local fluidics in confined chiral and achiral nematics. Ph.D. thesis, Faculty of mathematics and physics, University of Ljubljana, Slovenia, 2014
12. B. Jerome, Surface effects and anchoring in liquid crystals. Rep. Prog. Phys. **54**, 391 (1991)
13. G. Volovik, O. Lavrentovich, Topological dynamics of defects: boojums in nematic drops. Zh. Eksp. Teor. Fiz. **85**, 1997–2010 (1983)
14. A. Rapini, M. Papoular, Distorsion d'une lamelle nématique sous champ magnétique conditions d'ancrage aux parois. J. Phys. Colloq. **30**, C4–54 (1969)
15. M. Nobili, G. Durand, Disorientation-induced disordering at a nematic-liquid-crystal-solid interface. Phys. Rev. A **46**, R6174 (1992)
16. O. Lavrentovich, Topological defects in dispersed liquid crystals, or words and worlds around liquid crystal drops. Liq. Cryst. **24**, 117–126 (1998)

17. J.K. Gupta, J.S. Zimmerman, J.J. de Pablo, F. Caruso, N.L. Abbott, Characterization of adsorbate-induced ordering transitions of liquid crystals within monodisperse droplets. Langmuir **25**, 9016–9024 (2009)

18. M. Humar, I. Muševiš, Surfactant sensing based on whispering-gallery-mode lasing in liquid-crystal microdroplets. Opt. Express **19**, 19836–19844 (2011)

19. X. Wang, D.S. Miller, E. Bukusoglu, J.J. de Pablo, N.L. Abbott, *Topological defects in liquid crystals as templates for molecular self-assembly* (Nat, Mater, 2015)

20. H.-R. Trebin, The topology of non-uniform media in condensed matter physics. Adv. Phys. **31**, 195–254 (1982)

21. M.I. Monastyrsky, *Riemann, Topology, and Physics*, 2nd edn. (Birkhäuser, Boston, 2008)

22. G.P. Alexander, B.G.-G. Chen, E.A. Matsumoto, R.D. Kamien, Colloquium: disclination loops, point defects, and all that in nematic liquid crystals. Rev. Mod. Phys. **84**, 497–514 (2012)

23. S. Čopar, S. Žumer, Quaternions and hybrid nematic disclinations. Proc. R. Soc. A **469**, 20130204 (2013)

24. M. Nikkhou et al., Light-controlled topological charge in a nematic liquid crystal. Nat. Phys. **11**, 183–187 (2015)

25. S. Čopar, Private communication

26. S. Čopar, S. Žumer, Topological and geometric decomposition of nematic textures. Phys. Rev. E **85**, 031701 (2012)

27. N. Madhusudana, R. Pratibha, Studies on high strength defects in nematic liquid crystals? Mol. Cryst. Liq. Cryst. **103**, 31–47 (1983)

28. O. Lavrentovich, Y.A. Nastishin, Defects in degenerate hybrid aligned nematic liquid crystals. Europhys. Lett. **12**, 135 (1990)

29. O. Lavrentovich, V. Pergamenshchik, Patterns in thin liquid crystal films and the divergence ("surfacelike") elasticity. Int. J. Mod. Phys. B **9**, 2389–2437 (1995)

30. M. Kléman, O.D. Lavrentovich, Topological point defects in nematic liquid crystals. Philos. Mag. **86**, 4117–4137 (2006)

31. H. Brezis, J.-M. Coron, E.H. Lieb, Harmonic maps with defects. Commun. Math. Phys. **107**, 649–705 (1986)

32. L. Giomi, Ž. Kos, M. Ravnik, A. Sengupta, Dynamical and topological singularities cross-talk in flowing nematic liquid crystals. Proc. Natl. Acad. Sci. U. S. A. **114**(29), E5771–E5777 (2017)

33. T. Lubensky, D. Pettey, N. Currier, H. Stark, Topological defects and interactions in nematic emulsions. Phys. Rev. E **57**, 610 (1998)

34. P. Poulin, D. Weitz, Inverted and multiple nematic emulsions. Phys. Rev. E **57**, 626–637 (1998)

35. M. Škarabot et al., Interactions of quadrupolar nematic colloids. Phys. Rev. E **77**, 031705 (2008)

36. T. Machon, G.P. Alexander, Global defect topology in nematic liquid crystals. Proc. R. Soc. A **472**, 20160265 (2016)

37. T. Machon, G.P. Alexander, Knots and nonorientable surfaces in chiral nematics. Proc. Natl. Acad. Sci. USA **110**, 14174–14179 (2013)

38. D. Seč, S. Čopar, S. Žumer, Topological zoo of free-standing knots in confined chiral nematic fluids. Nat. Commun. **5**, 3057 (2014)

39. K. Jänich, Topological properties of ordinary nematics in 3-space. Acta Appl. Math. **8**, 65–74 (1987)

40. M. Yada, J. Yamamoto, H. Yokoyama, Direct observation of anisotropic interparticle forces in nematic colloids with optical tweezers. Phys. Rev. Lett. **92**, 185501 (2004)

41. P. Poulin, H. Stark, T. Lubensky, D. Weitz, Novel colloidal interactions in anisotropic fluids. Science **275**, 1770–1773 (1997)

42. H. Stark, Physics of colloidal dispersions in nematic liquid crystals. Phys. Rep. **351**, 387–474 (2001)

43. K. Takahashi, M. Ichikawa, Y. Kimura, Direct measurement of force between colloidal particles in a nematic liquid crystal. J. Phys.: Condens. Matter **20**, 075106 (2008)

44. T.W. Kibble, Topology of cosmic domains and strings. J. Phys. A **9**, 1387–1398 (1976)

45. W.H. Zurek, Cosmological experiments in condensed matter systems. Phys. Rep. **276**, 177–221 (1996)
46. I. Chuang, R. Durrer, N. Turok, B. Yurke, Cosmology in the laboratory: defect dynamics in liquid crystals. Science **251**, 1336–1342 (1991)
47. M. Zapotocky, L. Ramos, P. Poulin, T. Lubensky, D. Weitz, Particle-stabilized defect gel in cholesteric liquid crystals. Science **283**, 209–212 (1999)
48. B. Senyuk et al., Topological colloids. Nature **493**, 200–205 (2012)
49. U. Tkalec, M. Ravnik, S. Čopar, S. Žumer, I. Muševiš, Reconfigurable knots and links in chiral nematic colloids. Science **333**, 62–65 (2011)
50. A. Martinez et al., Mutually tangled colloidal knots and induced defect loops in nematic fields. Nat. Mater. **13**, 258–263 (2014)
51. L. Tran et al., Lassoing saddle splay and the geometrical control of topological defects. Proc. Natl. Acad. Sci. USA **113**, 7106–7111 (2016)
52. C. Williams, P. Piéranski, P. Cladis, Nonsingular s = + 1 screw disclination lines in nematics. Phys. Rev. Lett. **29**, 90 (1972)
53. W.E. Haas, J.E. Adams, New optical storage mode in liquid crystals. Appl. Phys. Lett. **25**, 535–537 (1974)
54. M. Kawachi, O. Kogure, Y. Kato, Bubble domain texture of a liquid crystal. Jpn. J. Appl. Phys. **13**, 1457 (1974)
55. S. Pirkl, P. Ribiere, P. Oswald, Forming process and stability of bubble domains in dielectrically positive cholesteric liquid crystals. Liq. Cryst. **13**, 413–425 (1993)
56. I.I. Smalyukh, Y. Lansac, N.A. Clark, R.P. Trivedi, Three-dimensional structure and multistable optical switching of triple-twisted particle-like excitations in anisotropic fluids. Nat. Mater. **9**, 139–145 (2010)
57. J.D. Jackson, *Electrodynamics* (Wiley Online Library, 1975)
58. D.K. Yang, *Fundamentals of Liquid Crystal Devices* (John Wiley & Sons, 2014)
59. P. Yeh, C. Gu, *Optics of Liquid Crystal Displays* (John Wiley & Sons, 1999)
60. D.W. Berreman, Optics in stratified and anisotropic media - 4 × 4-matrix formulation. J. Opt. Soc. Am. **62**, 502–510 (1972)
61. I. Smalyukh, O. Lavrentovich, Three-dimensional director structures of defects in Grandjean-Cano wedges of cholesteric liquid crystals studied by fluorescence confocal polarizing microscopy. Phys. Rev. E **66**, 051703 (2002)
62. S. Shiyanovskii, I. Smalyukh, O. Lavrentovich, Computer simulations and fluorescence confocal polarizing microscopy of structures in cholesteric liquid crystals, in *Defects in Liquid Crystals: Computer Simulations, Theory and Experiments* (Springer, 2001), pp. 229–270

Chapter 3
Liquid Crystal Droplets

Droplets of liquid crystals are a relatively well studied system because of the ease of their preparation and the richness of phenomena which can be observed in them. The simplest way of preparing them is to mix a small amount of a LC with a medium which will not dissolve the LC, for example an organic-molecule-based liquid crystal with a polar solvent such as water. Droplets can also be formed with phase separation as in polymer-dispersed LC [1], or with microfluidics which gives substantial control over their sizes [2, 3]. Droplets of various LC phases have been studied so far: smectic [4, 5], columnar [6] and blue phases [7–9], but by far the most frequently studied are nematic and cholesteric droplets.

The shape of the droplets depends on the interplay of surface tension and elastic forces in the droplet—if the surface tension is relatively large, the droplets are spherical, minimising the surface of the LC volume, but other shapes are also possible. If the surface energy is relatively small, the elastic distortions can cause the surface of the LC to undulate [10, 11]. The competition between elastic forces and surface tension can also lead to different anisotropic shapes of droplets. One example are elongated, spindle-shaped droplets called tactoids [12–14]. In cases where the surface energy is lowered with a surfactant, the interplay of surface energy and elastic constants close to a smectic phase transition can drive the formation of extended tube-like structures, which can either be stabilised [15, 16] or even divide into smaller droplets [17]. Droplets with holes can be formed either in stress-yield mediums [18] or by confining a LC in handlebody shaped polymer cavities [19]. Thin spherical LC shells with planar anchoring have been researched extensively both theoretically and experimentally [20]. The Poincaré-Hopf theorem, which connects the sum of the winding numbers of 2D defects on a closed surface to its genus, has been verified for LC shells numerically [21, 22] and experimentally [23, 24].

If a nematic liquid crystal is used, the equilibrium director structure in the droplet predominantly depends on the anchoring of the molecules on the surface of the droplet. For a spherical nematic droplet with planar anchoring the Poincaré-Hopf theorem states that the sum of winding numbers of defects must be equal to 2, which means 2 surface boojums with $k = 1$ must be present [4, 25]. This usually results in a bipolar structure (Fig. 3.1a). With homeotropic anchoring, a radial structure with a

© Springer Nature Switzerland AG 2018   29
G. Posnjak, *Topological Formations in Chiral Nematic Droplets*,
Springer Theses, https://doi.org/10.1007/978-3-319-98261-8_3

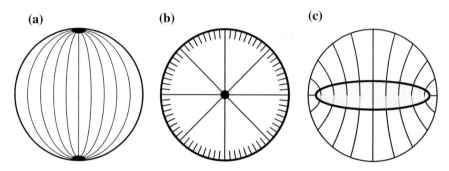

Fig. 3.1 Director structures in nematic droplets. **a** In droplets with planar anchoring a bipolar structure with two boojums is most the common. Nematic droplets with homeotropic anchoring have either **b** a radial structure with a central radial hedgehog or **c** a circumferential disclination ring

single point defect in the bulk LC is usually the most stable (Fig. 3.1b) but variation of anchoring strength and elastic constants of the LC can force the point defect to expand into a closed disclination loop running circumferentially around the droplet (Fig. 3.1c) [25, 26].

Because the structure in a nematic droplet is so sensitive to the anchoring conditions on the droplet surface, droplets can be used for detecting the presence of different chemicals. For example, small amounts of surfactants which are composed of hydrophobic and hydrophilic parts can change the anchoring in aqueous dispersions of LC from planar to homeotropic [27, 28]. Alternatively, tiny amounts of biological molecules can alter the structure by being concentrated in the topological defects [29].

In a droplet with homeotropic anchoring, the director on the surface is parallel to the normal of the surface. This means that the total topological charge inside the droplet must be $q = +1$ according to the Gauss-Bonnet theorem [Eq. (2.17)]. In a nematic droplet this usually means that in equilibrium only one radial point defect is present. If additional defects are formed, for example in a quench, they are annihilated when the director relaxes its elastic deformations and the total topological charge of the equilibrated state is again equal to $+1$. Additional defects can be stabilised if some kind of a barrier is inserted to prevent the annihilation of surplus defects. For example, colloidal particles with homeotropic anchoring can be added to the droplet, each carrying $+1$ topological charge. If only a single particle is present, it will occupy the centre of the droplet and no additional defect will form. If more particles are added, additional -1 defects will form, so that the total topological charge, determined by the homeotropic anchoring on the surface of the droplet will be compensated to $+1$ [30, 31].

A specific property of droplets compared to other confined systems is their symmetry—a structure can be embedded in a spherical confinement in any orientation, giving rise to several possible different appearances of the same structure.

Caution is even more critical in cholesteric droplets, where often several metastable states are possible and the nontrivial optics of the system hinders interpretation of optical textures [32–37].

3.1 Chiral Nematic Droplets

Chiral nematic droplets have been studied extensively: their formation in polymer-dispersed LC [1], the stablity of structures [34, 35, 38–45], structural transitions between them [46], influence of external fields on the structure [38, 45, 47–49], insertion of fibers [50] and assembly of nanoparticles in surface or bulk defects [51–53].

Chiral nematic droplets have an additional free parameter compared to nematic droplets—the chiral pitch p. The twisting of the director field can be embedded into the spherical confinement of a droplet in many different ways and therefore several different director structures with different elastic energies can emerge for a given set of parameters. Structures with non-minimal energy can be stable because of energy barriers associated with unwinding of the twisted structure. Because of their stability and non-minimal energy they are metastable. The main parameter determining the energy and consequently the stability of different structures is the diameter to pitch ratio $N = 2d/p_0$ [34, 44], where d is the droplet diameter and p_0 is the intrinsic pitch of the mixture—the pitch of the cholesteric layers of the LC/chiral dopant mixture in equilibrated bulk. The pitch of a layered structure in confined space such as a droplet or a wedge cell can be different from this equilibrium value because the structure can be elastically deformed to satisfy the boundary conditions. The factor of 2 in the formula means that the relative chirality parameter N counts the number of π twists the director could perform across the diameter of the droplet.

Chiral nematic droplets show a large variety of possible structures [34, 35, 38–44]. With planar anchoring and N greater than about 7, the most stable structure is onion-like with concentric spherical layers with either a diametral (Fig. 3.2a) or radial discontinuity in the cholesteric layers (Fig. 3.2b) connecting to two surface boojums [34]. In the diametric spherical structure (DSS), the two boojums are located on opposite poles of the droplet and in the radial spherical structure (RSS), they are positioned next to each other on the surface of the droplet. If the pitch of the layers in such a structure is comparable to the wavelength of light in the LC medium, the periodicity of the structure gives rise to selective Bragg reflections as discussed in Sect. 2.6. This property has been utilised in several applications, for example as polymer-dispersed LC for temperature-tunable paints [54, 55] and smart windows [1, 56], spherical resonator cavities in paintable [57, 58] and 3D lasers [59, 60] and holographic coding [61, 62].

If the diameter to pitch ratio is reduced, other structures become stable. A structure with two boojums and coaxial, cylindrical layers, simply called bipolar structure (BS), is shown in Fig. 3.2c. A similar planar bipolar structure (PBS) also has two diametrically positioned boojums, but the nested layers are flattened as shown in

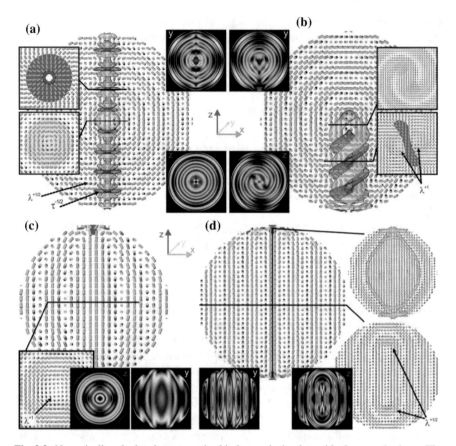

Fig. 3.2 Numerically calculated structures in chiral nematic droplets with planar anchoring. **a** The diametric spherical structure (DSS) in a droplet with $N = 10$ in which the diametric $+1$ disclination dissociates into a series of $\tau^{-1/2}$ and $\lambda^{+1/2}$ rings. **b** The radial spherical structure (RSS) in a droplet with $N = 10$, where two λ^{+1} disclinations helically wrap around each other and end in a pair of boojums. **c** The bipolar structure (BS) in a droplet with $N = 6$ with coaxial cylindrical layers and two boojums. **d** The planar bipolar structure (PBS) in a droplet with $N = 8$ with flattened coaxial cylindrical layers and two boojums. The insets show cross-sections of the structures and simulated transmission images under crossed polarisers looking either along or perpendicular to the symmetry axis of the droplet. Singular regions are shown in red color. Reproduced from Ref. [34] with permission of The Royal Society of Chemistry

Fig. 3.2d. In planar droplets with $N < 2$, a structure similar to the BS is found, but without real layering—it is called the double-twist structure [35]. Two more exotic structures were found in numerical simulations at moderate N (between 4 and 7), but have not yet been demonstrated experimentally. These are the closely related Lyre (Fig. 3.3a) and Yeti structures (Fig. 3.3b), in which the deformations of the cholesteric layers are spread throughout the droplet [34].

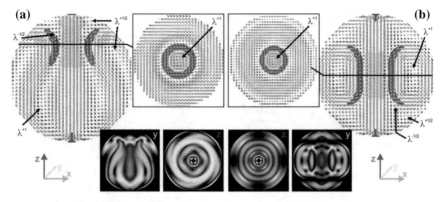

Fig. 3.3 Two exotic numerically calculated structures in chiral nematic droplets with planar anchoring. The Lyre (**a**) in a droplet with $N = 4$ and the Yeti (**b**) in a droplet with $N = 5$. Reproduced from Ref. [34] with permission of The Royal Society of Chemistry

3.2 Chiral Nematic Droplets with Homeotropic Anchoring

Chiral nematic droplets with homeotropic anchoring (CNDHO) have not been studied extensively prior to this study. Only a few experimental studies were conducted, which found that for large N the equilibrium structures of CNDHO appear similar to the ones in cholesteric droplets with planar anchoring and for $N < 2$ they found radial configurations with a single point defect in the centre or structures with circumferential ring defects, similarly as in nematic droplets with homeotropic anchoring [38, 39, 47].

A recent numerical study of this system [44] found that the twisting chiral structure in CNDHO is strongly frustrated because it does not match the radial ordering of the molecules on the surface of the droplet. This frustration prevents the cholesteric layers to fully relax, leading to extended line defects which terminate the cholesteric layers at the surface of the droplet. An example of this is a structure with parallel layers with a line defect appearing where the orientation of director at the edge of the layers does not match the anchoring on the surface, resulting in a spiralling line defect along the surface of the droplet as shown in Fig. 3.4.

The numerical experiments found that the frustration of cholesteric ordering by the homeotropic anchoring gives rise to a rich array of possible metastable states at a given N [44]. These states can be reached randomly by temperature quenches: if such a droplet is heated to isotropic phase and quickly cooled back to the chiral nematic phase, the cholesteric layers do not have sufficient time to relax to a parallel configuration, resulting in a dense tangle of line defects. After the structure is thermalised by elastic relaxation of the deformations, some of these defects can be trapped between the cholesteric layers inside the droplet and cannot be expelled to the surface as shown in Fig. 3.5. Because of this, the line defects can be arranged in complex structures and the study found that a line defect can occasionally form a knot, or

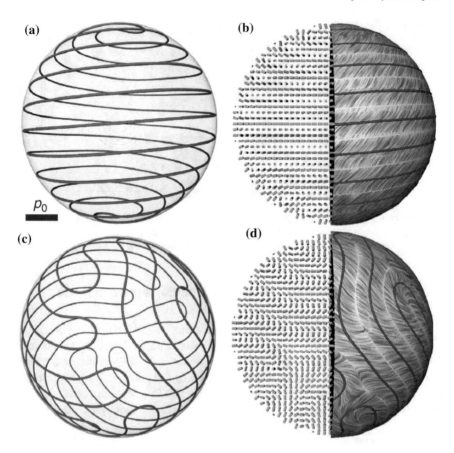

Fig. 3.4 Examples of numerically predicted structures in CNDHO. **a, b** A stable structure at $N = 12$. A defect line with $k = -1/2$ winding runs along the surface in a helical fashion. The left side of (**b**) shows the layer-like ordering of the director in the bulk of the droplet and the right side the ordering on the surface of the droplet. **c, d** A metastable structure at $N = 12$, with disordered layers in the bulk. Reprinted by permission from Macmillan Publishers Ltd: Nature Communications [44], copyright 2014

two or more closed line defects can be interwoven to form various links [44]. Such formations are not equilibrium structures with minimal free energy. This is because the additional defect volume, which is needed to tie the knots, increases the energy of the structure significantly, but the energy barriers associated with disentanglement of the structures are high enough to render the knots and links metastable [44].

So far knots and links in physical fields have been only observed either in dynamic systems or stabilised with additional constraints. Examples of knots in dynamic systems are zero amplitude field lines in electromagnetic fields [63–67] and knotted vortices in liquids [68]. In nematic fields, knots of line defects can be stabilised with colloidal inclusions, which serve as barriers preventing the topological defects from

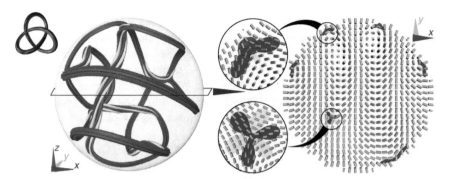

Fig. 3.5 An example of a numerically predicted structure in CNDHO at $N = 5$ with a knoted defect line. The left image shows how profile of the $k = -1/2$ defect line twists as it runs through the volume of the droplet. In this case the defect line forms a trefoil knot as shown in the top left inset. The right image shows how the defect line is trapped between the cholesteric layers in a cross-section plane of the droplet. The two insets show details of the director profile around the disclination line. Reprinted by permission from Macmillan Publishers Ltd: Nature Communications [44], copyright 2014

annihilating [69–71]. Another example of stabilisation of a physical knot is a knot soliton in the order parameter of a spinor Bose-Einstein condensate, stabilised by an external inhomogeneous magnetic field [72].

A knot in droplet would be a static non-trivial topological structure, stabilised by the frustrating confinement and the twisted ordering of the material itself without any additional constraints. This could enable non-trivial, topologically protected structures to be easily reconfigurable by manipulating the material properties, for example by using photosensitive chiral dopants [73] to change the relative chirality of the droplet and with this alter the stability of the knotted or linked defect structures.

An experimental report on the different textures which appear in chiral nematic droplets with homeotropic anchoring was published during our study [45]. At large N the study found spiralling defect lines on the surfaces of droplets and layered structures in the bulk of the droplets. It also suggested occurrence of different structures at moderate N by presenting various polarised and non-polarised wide-field optical textures of the droplets, but no attempts of reconstruction and characterisation of the structures were made. Control over the optical texture of a cholesteric droplet was demonstrated with temperature quenches, electric field and photo-activated chirality switching but as only wide field microscopy was used in the study, the effect on the structures was not analysed.

Another study published during our research conducted FCPM imaging on CNDHO and attempted to reconstruct the director field from the images at different polarisations [74]. The study used relatively large droplets ($\approx$50–100 μm) with long pitches ($p \approx 90$ μm) and relatively high birefringence (MLC-7026 liquid crystal, $\Delta n = 0.082$ [75]), resulting in a Mauguin parameter discussed in Sect. 2.6.1 well in the wave-guiding regime ($0.5 \, p \, \Delta n = 3.7$ μm $\gg \lambda$). The study took the out-of-plane tilting of director into account only schematically, leading to debatable interpretations of the structures. They concluded that the structures in the studied droplets have line defects with $+1$ winding and therefore the total topological charge is even, in contradiction to the numerical study by Seč et al. [44], which predicted the line defects to have half-integer winding and therefore odd topological charge. The even topological charge fundamentally disagrees with the homeotropic anchoring of the droplet, which due to Gauss-Bonnet theorem leads to a $+1$ total charge in a droplet. This disagreement casts doubts on the validity of the reconstruction procedure presented in the study.

The search for the knotted and linked defect states from Ref. [44] served as a motivation for our study but the method of reconstructing the director field from FCPM data presented in Chap. 6 enabled us to conduct a systematic study of various metastable states found in CNDHO. We will see, that in contrast to the numerical study, point defects are the predominant form of singular regions which we found in CNDHO. We will discuss the differences between our experimental and the numerical study which could lead to this discrepancy in the discussion at the end of the Thesis.

References

1. J. Doane, A. Golemme, J.L. West, J. Whitehead Jr., B.-G. Wu, Polymer dispersed liquid crystals for display application. Mol. Cryst. Liq. Cryst. **165**, 511–532 (1988)
2. A. Utada et al., Monodisperse double emulsions generated from a microcapillary device. Science **308**, 537–541 (2005)
3. R.K. Shah et al., Designer emulsions using microfluidics. Mater. Today **11**, 18–27 (2008)
4. O. Lavrentovich, Topological defects in dispersed liquid crystals, or words and worlds around liquid crystal drops. Liq. Cryst. **24**, 117–126 (1998)
5. J. Jeong, M.W. Kim, Confinement-induced transition of topological defects in smectic liquid crystals: From a point to a line and pearls. Phys. Rev. Lett. **108**, 207802 (2012)
6. J. Jeong, Z.S. Davidson, P.J. Collings, T.C. Lubensky, A.G. Yodh, Chiral symmetry breaking and surface faceting in chromonic liquid crystal droplets with giant elastic anisotropy. Proc. Natl. Acad. Sci. USA **111**, 1742–7 (2014)
7. E. Kemiklioglu, L.-C. Chien, Polymer-encapsulated blue phase liquid crystal droplets. Appl. Phys. Express **7**, 091701 (2014)
8. J.A. Martínez-González et al., Blue-phase liquid crystal droplets. Proc. Natl. Acad. Sci. USA **112**, 13195–13200 (2015)
9. E. Bukusoglu, X. Wang, J.A. Martinez-Gonzalez, J.J. de Pablo, N.L. Abbott, Stimuli-responsive cubosomes formed from blue phase liquid crystals. Adv. Mater. **27**, 6892–6898 (2015)
10. H. Yokoyama, S. Kobayashi, H. Kamei, Deformations of a planar nematic-isotropic interface in uniform and nonuniform electric fields. Mol. Cryst. Liq. Cryst. **129**, 109–126 (1985)
11. J. Yoshioka et al., Director/barycentric rotation in cholesteric droplets under temperature gradient. Soft Matter (2014)
12. J. Bernal, I. Fankuchen, X-ray and crystallographic studies of plant virus preparations: I. introduction and preparation of specimens ii. modes of aggregation of the virus particles. J. Gen. Physiol. **25**, 111 (1941)
13. L. Tortora, O.D. Lavrentovich, Chiral symmetry breaking by spatial confinement in tactoidal droplets of lyotropic chromonic liquid crystals. Proc. Natl. Acad. Sci. USA **108**, 5163–5168 (2011)
14. V. Jamali et al., Experimental realization of crossover in shape and director field of nematic tactoids. Phys. Rev. E **91**, 042507 (2015)

15. K. Peddireddy, P. Kumar, S. Thutupalli, S. Herminghaus, C. Bahr, Solubilization of thermotropic liquid crystal compounds in aqueous surfactant solutions. Langmuir **28**, 12426–12431 (2012)
16. K. Peddireddy, P. Kumar, S. Thutupalli, S. Herminghaus, C. Bahr, Myelin structures formed by thermotropic smectic liquid crystals. Langmuir **29**, 15682–15688 (2013)
17. O. Lavrentovich, Y.A. Nastishin, Division of drops of a liquid-crystal in the case of a cholesteric-smectic-A phase-transition. J. Exp. Theor. Phys. Lett. **40**, 1015–1019 (1984)
18. E. Pairam et al., Stable nematic droplets with handles. Proc. Natl. Acad. Sci. USA **110**, 9295–9300 (2013)
19. M. Tasinkevych, M.G. Campbell, I.I. Smalyukh, Splitting, linking, knotting, and solitonic escape of topological defects in nematic drops with handles. Proc. Natl. Acad. Sci. USA **111**, 16268–16273 (2014)
20. L. Mirantsev, E. de Oliveira, I. de Oliveira, M. Lyra, Defect structures in nematic liquid crystal shells of different shapes. Liq. Cryst. Rev. **4**, 35–58 (2016)
21. S. Kralj, R. Rosso, E.G. Virga, Curvature control of valence on nematic shells. Soft Matter **7**, 670–683 (2011)
22. T.-S. Nguyen, J. Geng, R.L. Selinger, J.V. Selinger, Nematic order on a deformable vesicle: theory and simulation. Soft Matter **9**, 8314–8326 (2013)
23. T. Lopez-Leon, V. Koning, K. Devaiah, V. Vitelli, A. Fernández-Nieves, Frustrated nematic order in spherical geometries. Nat. Phys. **7**, 391–394 (2011)
24. D. Seč et al., Defect trajectories in nematic shells: role of elastic anisotropy and thickness heterogeneity. Phys. Rev. E **86**, 020705 (2012)
25. G. Volovik, O. Lavrentovich, Topological dynamics of defects: boojums in nematic drops. Zh. Eksp. Teor. Fiz. **85**, 1997–2010 (1983)
26. J.K. Gupta, J.S. Zimmerman, J.J. de Pablo, F. Caruso, N.L. Abbott, Characterization of adsorbate-induced ordering transitions of liquid crystals within monodisperse droplets. Langmuir **25**, 9016–9024 (2009)
27. S. Sivakumar, K.L. Wark, J.K. Gupta, N.L. Abbott, F. Caruso, Liquid crystal emulsions as the basis of biological sensors for the optical detection of bacteria and viruses. Adv. Funct. Mater. **19**, 2260–2265 (2009)
28. M. Humar, I. Muševič, Surfactant sensing based on whispering-gallery-mode lasing in liquid-crystal microdroplets. Opt. Express **19**, 19836–19844 (2011)
29. I.-H. Lin et al., Endotoxin-induced structural transformations in liquid crystalline droplets. Science **332**, 1297–1300 (2011)
30. P. Poulin, H. Stark, T. Lubensky, D. Weitz, Novel colloidal interactions in anisotropic fluids. Science **275**, 1770–1773 (1997)
31. P. Poulin, D. Weitz, Inverted and multiple nematic emulsions. Phys. Rev. E **57**, 626–637 (1998)
32. J. Ding, Y. Yang, Birefringence patterns of nematic droplets. Jpn. J. Appl. Phys. **31**, 2837 (1992)
33. P.S. Drzaic, A case of mistaken identity: spontaneous formation of twisted bipolar droplets from achiral nematic materials. Liq. Cryst. **26**, 623–627 (1999)
34. D. Seč, T. Porenta, M. Ravnik, S. Žumer, Geometrical frustration of chiral ordering in cholesteric droplets. Soft Matter **8**, 11982–11988 (2012). https://dx.doi.org/10.1039/C2SM27048
35. J. Yoshioka, F. Ito, Y. Tabe, Stability of double twisted structure in spherical cholesteric droplets. Soft Matter (2016)
36. G. Posnjak, S. Čopar, I. Muševič, Points, skyrmions and torons in chiral nematic droplets. Sci. Rep. **6**, 26361 (2016)
37. U. Mur et al., Ray optics simulations of polarised microscopy textures in chiral nematic droplets. Liq. Cryst. **44**, 679–687 (2017)
38. S. Candau, P. Le Roy, F. Debeauvais, Magnetic field effects in nematic and cholesteric droplets suspended in a isotropic liquid. Mol. Cryst. Liq. Cryst. **23**, 283–297 (1973)
39. M. Kurik, O. Lavrentovich, Negative-positive monopole transitions in cholesteric liquid crystals. J. Exp. Theor. Phys. Lett. **35**, 444–447 (1982)

40. M. Kurik, O. Lavrentovich, Topological defects of cholesteric liquid crystals for volumes with spherical shape. Mol. Cryst. Liq. Cryst. **72**, 239–246 (1982)
41. Y. Bouligand, F. Livolant, The organization of cholesteric spherulites. J. Phys. **45**, 1899–1923 (1984)
42. J. Bezić, S. Žumer, Structures of the cholesteric liquid crystal droplets with parallel surface anchoring. Liq. Cryst. **11**, 593–619 (1992)
43. F. Xu, P. Crooker, Chiral nematic droplets with parallel surface anchoring. Phys. Rev. E **56**, 6853 (1997)
44. D. Seč, S. Žumer, Topological zoo of free-standing knots in confined chiral nematic fluids. Nat. Commun. **5**, 3057 (2014). https://www.nature.com/ncomms/
45. T. Orlova, S.J. Asshoff, T. Yamaguchi, N. Katsonis, E. Brasselet, Creation and manipulation of topological states in chiral nematic microspheres. Nat. Commun. **6**, 7603 (2015)
46. Y. Zhou et al., Structural transitions in cholesteric liquid crystal droplets. ACS Nano **10**, 6484–6490 (2016)
47. H.-S. Kitzerow, P. Crooker, Electric field effects on the droplet structure in polymer dispersed cholesteric liquid crystals. Liq. Cryst. **13**, 31–43 (1993)
48. J. Bajc, J. Bezić, S. Žumer, Chiral nematic droplets with tangential anchoring and negative dielectric anisotropy in an electric field. Phys. Rev. E **51**, 2176 (1995)
49. J. Bajc, S. Zumer, Structural transition in chiral nematic liquid crystal droplets in an electric field. Phys. Rev. E **55**, 2925 (1997)
50. Y. Geng et al., Liquid crystal necklaces: cholesteric drops threaded by thin cellulose fibres. Soft Matter **9**, 7928–7933 (2013)
51. M. Rahimi et al., Nanoparticle self-assembly at the interface of liquid crystal droplets. Proc. Natl. Acad. Sci. USA **112**, 5297–5302 (2015)
52. Y. Li et al., Colloidal cholesteric liquid crystal in spherical confinement. Nat. Commun. **7** (2016)
53. Y. Li et al., Periodic assembly of nanoparticle arrays in disclinations of cholesteric liquid crystals. Proc. Natl. Acad. Sci. USA **114**, 2137–2142 (2017)
54. C. Smith, D. Sabatino, T. Praisner, Temperature sensing with thermochromic liquid crystals. Exp. Fluids **30**, 190–201 (2001)
55. I. Sage, Thermochromic liquid crystals. Liq. Cryst. **38**, 1551–1561 (2011)
56. D.-K. Yang, L.-C. Chien, J. Doane, Cholesteric liquid crystal/polymer dispersion for haze-free light shutters. Appl. Phys. Lett. **60**, 3102–3104 (1992)
57. D.J. Gardiner et al., Paintable band-edge liquid crystal lasers. Opt. Express **19**, 2432–2439 (2011)
58. P. Hands et al., Band-edge and random lasing in paintable liquid crystal emulsions. Appl. Phys. Lett. **98**, 141102 (2011)
59. M. Humar, I. Muševič, 3D microlasers from self-assembled cholesteric liquid-crystal microdroplets. Opt. Express **18**, 26995–27003 (2010)
60. M. Humar, Liquid-crystal-droplet optical microcavities. Liq. Cryst. **43**, 1937–1950 (2016)
61. S.J. Asshoff et al., Superstructures of chiral nematic microspheres as all-optical switchable distributors of light. Sci. Rep.**5** (2015)
62. Y. Geng et al., High-fidelity spherical cholesteric liquid crystal Bragg reflectors generating unclonable patterns for secure authentication. Sci. Rep. **6** (2016)
63. A.F. Ranada, J.L. Trueba, Ball lightning an electromagnetic knot? Nature **383**, 32–32 (1996)
64. L. Faddeev, A.J. Niemi, Stable knot-like structures in classical field-theory. Nature **387**, 58–61 (1997)
65. W.T. Irvine, D. Bouwmeester, Linked and knotted beams of light. Nat. Phys. **4**, 716–720 (2008)
66. M.R. Dennis, R.P. King, B. Jack, K. O'Holleran, M.J. Padgett, Isolated optical vortex knots. Nat. Phys. **6**, 118–121 (2010)
67. H. Kedia, I. Bialynicki-Birula, D. Peralta-Salas, W.T. Irvine, Tying knots in light fields. Phys. Rev. Lett. **111**, 150404 (2013)
68. D. Kleckner, W.T. Irvine, Creation and dynamics of knotted vortices. Nat. Phys. **9**, 253–258 (2013)

69. U. Tkalec, M. Ravnik, S. Copar, S. žumer, I. & Muševič, Reconfigurable knots and links in chiral nematic colloids. Science **333**, 62–65 (2011)
70. T. Machon, G.P. Alexander, Knots and nonorientable surfaces in chiral nematics. Proc. Natl. Acad. Sci. USA **110**, 14174–14179 (2013)
71. A. Martinez et al., Mutually tangled colloidal knots and induced defect loops in nematic fields. Nat. Mater. **13**, 258–263 (2014)
72. D.S. Hall et al., Tying quantum knots. Nat. Phys. **12**, 478–483 (2016)
73. C. Ruslim, K. Ichimura, Conformational effect on macroscopic chirality modification of cholesteric mesophases by photochromic azobenzene dopants. J. Phys. Chem. B **104**, 6529–6535 (2000)
74. J.-K. Guo, J.-K. Song, Three-dimensional reconstruction of topological deformation in chiral nematic microspheres using fluorescence confocal polarizing microscopy. Opt. Express **24**, 7381–7386 (2016)
75. S.D. Kim, J.-K. Guo, J.-K. Song, Suspended, one-side anchored, or double-side anchored nematic droplets in an isotropic medium. Liq. Cryst. 1–7 (2016)

Chapter 4
Experimental Inspection of Director Fields

In this chapter we will first explain how optical microscopy can be used to reveal the director orientation in liquid crystals, then introduce confocal microscopy which allows us to scan the sample with 3D resolution and finally present fluorescent confocal polarising microscopy (FCPM) as an advanced tool for studying director orientation. In the last section we will derive the angular dependence of fluorescence intensities in FCPM and discuss what information we can extract from the experimental data.

In previous chapters we have seen that the liquid crystal director can form complex structures even in equilibrium. The most common approach to examining director field structures on the micrometre scale is polarised microscopy. A typical polarised microscopy set-up is shown in Fig. 4.1. If unpolarised light was used, the only contrast in the images would be due to variation of transmitted intensity. If the LC sample is a material with homogeneous composition, there is no considerable variation of absorption and intensity varies only because of scattering and lensing effects due to the variation of refractive index which depends on the director orientation.

More information can be extracted with polarised light under crossed polarisers which reveals director deviations from the direction of the polarisers. An additional lambda wave plate reveals information on the direction of the director deviation. Alternatively, parallel polarisers can be used, which reveal areas which strongly scatter light, such as defects. Usually a combination of the three methods is used to quickly examine a sample under a microscope to get a general idea about a structure and possibly identify the locations of topological defects. A more advanced version of polarised microscopy is called PolScope [1], which uses a variable retardation plate instead of a fixed lambda wave plate. By recording the transmitted intensity at different polarisations it measures the local retardation of the sample. This information can be used to calculate the director orientation in the image plane [2].

All these methods use wide-field transmission microscopy which means the transmitted light probes the director orientation in all the layers of the sample through which it passes. Because of this, wide-field microscopy methods only offer

© Springer Nature Switzerland AG 2018
G. Posnjak, *Topological Formations in Chiral Nematic Droplets*,
Springer Theses, https://doi.org/10.1007/978-3-319-98261-8_4

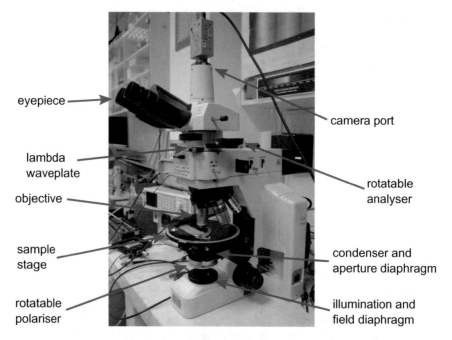

eyepiece

camera port

lambda
waveplate

objective

rotatable
analyser

sample
stage

condenser and
aperture diaphragm

rotatable
polariser

illumination and
field diaphragm

Fig. 4.1 A microscope used for polarised microscopy. The light coming from the bottom is polarised by a polariser and then focused on the sample by a condenser. The light that passes through the sample is collected by an objective and if selected, passed through a lambda waveplate and an analyser before coming to the eyepiece, where the magnified image of the sample can be observed. Alternatively, the light can be passed to a camera port for image recording. Both the polariser and analyser are rotatable in addition to the sample stage, so that any combination of polarisations and in-plane sample orientation can be selected

information about the director which is integrated through the whole sample thickness, yielding a projection of the sample to a 2D image. If the director field does not vary significantly along the optical axis of the microscope, this information can be sufficient to deduce the orientation of the LC molecules. In contrast, many features and details can be obscured in samples with z-dependent director field. Using high numerical aperture (NA) objectives with shallow depth of field and focusing at different depths can offer some insight into the third dimension of the sample by identifying which features are sharper at which focusing depth. The rest of the sample is still present as a blurred background, but if it is transparent, it does not hinder the imaging. Alternatively, if the out-of-focus sample is not transparent, the blurred background can obscure the image and not much depth resolution is gained by a high NA objective.

All physical measurements have some finite accuracy—the dimension of the probe. In the case of optical microscopes the probing dimension is a consequence of the wave nature of light. Light emitted from a point object cannot be imaged as a point due to the finite dimensions of the objective lens. Instead, its image has a radially symmetric modulated light intensity profile [3]. Two point objects can therefore be resolved only if they are separated at least by the distance between the central maximum and the first minimum of the intensity of their images. For incoherent light this distance is about half of the wavelength of the probing light, as stated by the Rayleigh criterion [4]:

$$\delta_{\text{Rayleigh}} = \frac{0.61\lambda}{n \sin\theta} , \tag{4.1}$$

where the numerical factor is due to the intensity profile, the factor λ/n is the wavelenght of the probing light in the material and $\sin\theta$ takes into account from how wide an angle we are observing the object. Both n and θ are connected to the objective used in the imaging, so the equation can be rewritten as

$$\delta_{\text{Rayleigh}} = \frac{0.61\lambda}{NA} , \tag{4.2}$$

where NA is the numerical aperture of the microscope lens.

4.1 Confocal Microscopy

If we are studying 3D samples, the three-dimensional image of a point source called a *point spread function* (PSF) becomes important. In the case of wide-field microscopy, the intensity from a point source only slowly drops along the optical axis, with the total light intensity in a perpendicular plane being constant [3]. This explains why blurred, out of focus objects are present in wide-field images. We can get around this problem by a clever design of the microscopy system: by inserting a series of lenses into the optical path (Fig. 4.2), the image of the sample is formed in intermediate planes which are confocal to the focusing plane. If we place an opaque disk with a small pinhole in the confocal plane between the sample and the final image, most of the light coming from the out-of-focus planes can be blocked as indicated in Fig. 4.2. A microscope using such a configuration is called a confocal microscope [3].

Figure 4.3 shows a comparison between the PSFs of a wide-field and a confocal microscope. The lateral size (Fig. 4.3a) of a confocal PSF is in the first approximation roughly the same as that of a wide-field microscope given in Eq. 4.2. The true advantage of a confocal over a wide-field microscope becomes apparent in the axial size of the PSF. In Fig. 4.3b we can see that the wide-field PSF is much larger than that of a confocal microscope. In the wide-field case, a blurred image of the point appears even when the focus is above or below the object. In the confocal case, the image of the point appears only in slices close to the object because of the rejection of the out-of-focus light by the pinhole.

The axial dimension of the confocal PSF is in the first approximation determined by the wavelength of the used light λ, the refractive index of the sample n and the numerical aperture of the objective NA [3]:

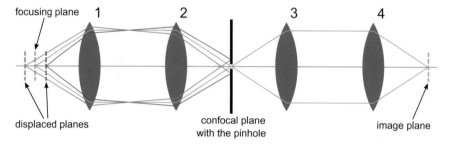

Fig. 4.2 A confocal microscope. The light coming from a plane, which is in focus (green beams) is collected by a lens (1) and transformed into a parallel beam. A subsequent lens (2) transforms this beam back to a convergent beam which is focused in a plane confocal to the focusing plane, and then again collected by a system of lenses (3) and (4) to form an image in the plane on the right side of the schematic. Beams originating from planes either deeper or shallower (orange and red beams) in the sample, are not parallel after passing the lens (1) and are not focused in the intermediate confocal plane. If a disk with a small pinhole is placed in this plane, such beams are blocked and therefore light from out-of-focus levels of sample won't be present in subsequent confocal planes

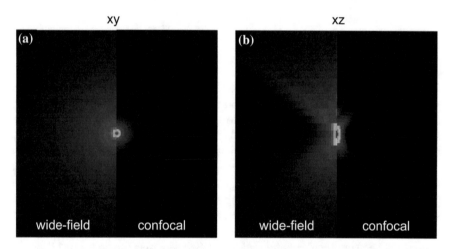

Fig. 4.3 A comparison of numerically calculated point spread functions of a wide-field and a confocal microscope. **a** The maximum intensity projection (MIP) of the PSF to the xy plane for a wide-field (left) and a confocal (right) microscope. The lateral dimensions of the central maximum (red centre) is approximately of the same size, but the wide-field PSF is spread-out in the whole xy plane whereas the confocal one is localised. **b** MIP of the two PSF to the xz plane. The intensity of the confocal PSF is localised around the sample plane, and in the wide-field case, much of the intensity is present in other planes. The PSF were calculated for an oil immersion objective with $NA = 1.4$, a pinhole size of 1 Airy disc and an index matched sample using the SVI calculator available at: https://svi.nl/NyquistCalculator (accessed 22. 3. 2017). The intensities are shown with a 1.7 gamma correction to highlight the dimmer regions

$$\delta_{\text{axial}} = \frac{1.26 n \lambda}{N A^2} , \tag{4.3}$$

This means that confocal microscopes are able to produce thin optical slices of the sample with a thickness on the order of λ. A whole 3D scan of the sample can be obtained by translating the focusing plane of the microscope through the sample and obtaining an image at each depth.

4.2 Fluorescent Confocal Polarising Microscopy

Confocal microscopes enable us to do 3D scans of samples, but do not intrinsically offer any information about the director field. To gain this information, we have to modify confocal microscopy by using polarised probing light and utilize a probing interaction with a directional dependence [5–7]. A common way to do this is to add fluorescent dyes to the LC sample which can be excited to a non-equilibrium molecular state with light of a certain wavelength and then relax back to the equilibrium state by emitting light of a longer wavelength (Fig. 4.4a). The wavelength of the emitted light corresponds to the energy level difference between the excited and ground state. The emitted (fluorescent) light has a longer wavelength than the excitation light and the difference in the location of the peak of excitation and emission spectrum is called the Stokes shift (Fig. 4.4b) [8].

To gain information about the orientation of the director field, the dye has to align in some way with respect to the director. The most commonly used regime of dye alignment is to use rigid, rod-like dye molecules with dipole moment along the axis of the molecule, which align parallel to the director of a calamitic LC [5, 7] but also other variations can be used [9]. In rod-like liquid crystals dye molecules with elongated shape can align with the local director and have a similar order parameter as the LC phase [10, 11].

The following derivation of the angular dependency of FCPM intensities builds on the discussions in the original literature on FCPM [7, 12–14] by taking into account the difference in the angular dependence on the polar tilt of the director with respect to the optical axis and its azimuthal orientation in the image plane. The derived results differ from the commonly cited equations for the angular dependence of FCPM intensity and allow for a more precise analysis of experimental data.

The angular dependence of fluorescence intensity in polarised confocal microscopy comes from two interactions—the excitation of the dye molecules and the emission of fluorescent light. The absorption of excitation light is the highest when the polarisation of the excitation light is aligned with the dipole moment of a dye molecule. It is enough for the dipole moment of the molecule to be only approximately aligned along the long axis of the molecule: the component of dipole moment which is perpendicular to the director, averages out because of nematic rotational symmetry around the director. Because of this, the average dipole moment is aligned with the director.

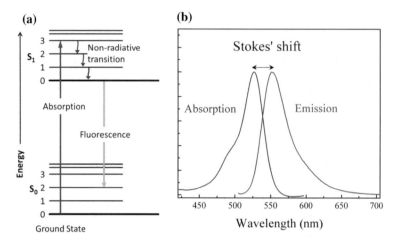

Fig. 4.4 Fluorescence of molecules. **a** An example of a Jablonski diagram of the molecular energy levels in a dye molecule. When a dye molecule absorbs a foton, it transitions from the ground state to one of the excited states. It relaxes back to the ground state through a series of transitions, where the one with the largest change in the energy levels corresponds to fluorescent emission. **b** An example of absorption and emission spectrum of the dye Rhodamine 6G in an unspecified medium. Emission wavelengths are Stokes shifted compared to absorption wavelengths because a part of the energy is dissipated in non-radiative transitions. Panel **a** reproduced from http://www.wikipedia.org/ under the terms of the Creative Commons CC0 1.0 Universal Public Domain Dedication (https://creativecommons.org/publicdomain/zero/1.0/), author Jacobkhed. Panel **b** is in public domain, reproduced from http://www.wikipedia.org/, author Sobarwiki

In the excitation stage, the excitation of the dipole moment of the dye molecules depends on the projection of the electrical field of excitation light on the direction of the director field. If we set a coordinate system so that the optical axis of the microscope is along z direction, we can write the director as

$$\mathbf{n} = (\cos \varphi \cos \theta, \sin \varphi \cos \theta, \sin \theta) , \tag{4.4}$$

where φ is the azimuthal angle measured from the x axis in the xy plane and θ the out-of-plane polar angle measured from the xy plane. For a fixed polarisation along x the projection of the electric field to the director is then:

$$E_{\text{proj}} = E_{\text{ex}} \cos \varphi \cos \theta . \tag{4.5}$$

The excitation of dye molecules depends on the energy provided by the pump beam, i.e. the light intensity, which means the excitation of the dye depends on the square of the electric field. If a nonlinear process is used for excitation such as two- or three-photon absorption or some other non-linear process that does not require the sample to be doped with a dye, the excitation becomes proportional to some power n of the excitation intensity [13]:

$$I_{ex} = (E_{ex} \cos \varphi \cos \theta)^{2n} . \tag{4.6}$$

The power n is the number of photons involved in the excitation: 1 for single-, 2 for two- and 3 for three-photon excitation. The emission intensity of the excited electrons in the dye molecule is proportional to I_{ex}, but because they emit as dipolar antennas oriented along the director field, the dependence of emitted intensity along the optical axis becomes

$$I_{det} \propto E_{ex}^{2n} \cos^{2n} \varphi \cos^{2n+2} \theta . \tag{4.7}$$

In the case of polarised detection a polariser aligned with the excitation polarisation is used before the detector and Eq. (4.7) has to be projected to the direction of the polariser, so the final expression for detected polarised intensity becomes:

$$I_{det,pol} \propto E_{ex}^{2n} \cos^{2n+2a} \varphi \cos^{2n+2} \theta , \tag{4.8}$$

with $a = 0$ for unpolarised and $a = 1$ for polarised detection.

In FCPM experiments with single photon excitation and polarised detection ($n = 1$, $a = 1$), the powers of both angle-dependent functions are 4, theoretically giving good angular sensitivity, with the fluorescence intensity being the highest when the dye is oriented along the excitation polarisation ($I_{\parallel}$) and zero when it is perpendicular to it ($I_{\perp}$). In experiments $I_{\perp}$ is not zero because dye dipoles have a non-zero component perpendicular to the director. Furthermore, if high NA objectives are used, the light beam is not a plane wave and can have a polarisation component out of the xy plane. The original references on FCPM [7, 12] report a contrast of $I_{\parallel} : I_{\perp} = 7.5$, but in our experiments in droplets with a setup described in Chap. 5 and an objective with $NA = 1.4$, a contrast of $I_{\parallel} : I_{\perp} \approx 3$ was regularly observed.

Because of the strong φ dependence, several excitation/detection polarisations have to be utilised in the experiment to evenly collect intensity over all azimuthal angles. Four polarisations separated by $\pi/4$ are sufficient to determine the orientation of the director projected to the xy plane. Intuitively this can be understood as follows: using two polarisations separated by 90° we can discern between orientations in one quadrant of the coordinate system, but not between α and $-\alpha$. With this we determine only one Stokes parameter, S_1 in Eq. (2.28). By using two additional polarisations at 45° and 135°, we can discriminate between α and $-\alpha$, or in other words determine the second Stokes parameter S_2 in Eq. (2.29). In this way we can exactly determine the angle α in the range $0° - 180°$. Because of nematic symmetry $\mathbf{n} = -\mathbf{n}$ this gives us perfect characterisation of the azimuthal angle φ.

By measuring the fluorescence at the four polarisations we collect intensities:

$$I_k = C \cos^{2n+2a}(\varphi - k\pi/4) \cos^{2n+2} \theta , \tag{4.9}$$

where k runs from 0 to 3 and C is a constant, dependent on the optical system, excitation intensity, quantum yield and the order parameter of the dye. The angles φ and θ can be obtained from I_k through the Stokes parameters [Eq. (2.27–2.29)] of

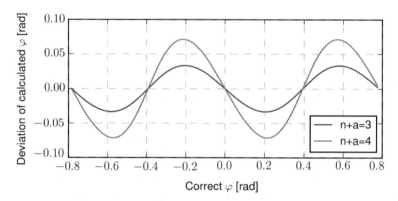

Fig. 4.5 Deviation of φ calculated from Eq. 4.14 from the correct value of φ, for $n + a > 2$

the fluorescent light [14]:

$$S_0 = I_{\text{tot}} = \frac{1}{2}(I_0 + I_1 + I_2 + I_3) \,, \qquad (4.10)$$

$$S_1 = I_0 - I_2 \,, \qquad (4.11)$$

$$S_2 = I_1 - I_3 \,. \qquad (4.12)$$

If we calculate the ratio S_2/S_1 and insert Eq. (4.9), we get

$$\frac{S_2}{S_1} = \frac{I_1 - I_3}{I_0 - I_2} = \frac{\cos^{2n+2a}(\varphi - \pi/4) - \cos^{2n+2a}(\varphi - 3\pi/4)}{\cos^{2n+2a}\varphi - \cos^{2n+2a}(\varphi - \pi/2)} = \tan 2\varphi \; f(\varphi) \,, \qquad (4.13)$$

where $f(\varphi)$ is a function, the form of which depends on $n + a$. For $n + a$ equal
to 1 or 2, it has a constant value equal to 1. For values of $n + a$ larger than 2, it
is a non-linear function which oscillates around 1, but can be taken as constant to
simplify the calculations in the first approximation. In that case, the angle φ can be
calculated from the relation:

$$\varphi = \frac{1}{2} \arctan \frac{I_1 - I_3}{I_0 - I_2} \,, \qquad (4.14)$$

which holds exactly for $n + a$ equal to 1 or 2 and approximately for higher values,
as can be seen in Fig. 4.5. The angle φ calculated from Eq. 4.14 lies in the $[0, \pi]$
range, which is sufficient for describing the director, because its projection to the xy
plane does not have an arrow and therefore angles in the $[\pi, 2\pi]$ range map to the
$[0, \pi]$ range.

To get the out-of-plane angle size θ, we have to find the size of projection of
director on the xy plane by summing the intensities at different polarisations to get
rid of the φ intensity dependence. Mathematically, in the case of polarised single-
or two-photon excitation and detection or three-photon excitation with unpolarised

detection, four polarisations separated by $\pi/4$ are sufficient because

$$\sum_{k=0}^{3} \cos^{2m}(\varphi - k\pi/4) \tag{4.15}$$

adds up to a constant for m values from 1 to 3. For $m = 4$ (three-photon excitation with polarised detection) this sum is not constant, but the azimuthal variation is less than 3% so in the first approximation it can be taken as constant. By inserting Eq. (4.9) into the expression for I_{tot}, we get

$$I_{tot} = \frac{C}{2}\cos^{2n+2}\theta \sum_{k=0}^{3} \cos^{2n+2a}(\varphi - k\pi/4) , \tag{4.16}$$

which can be simplified according to Eq. (4.15) to

$$I_{tot} \propto \cos^{2n+2}\theta , \tag{4.17}$$

which is exact for n equal to 1 or 2, and correct within 3% for $n = 3$. I_{tot} can be gained in a single measurement if circularly polarised or unpolarised light is used for excitation and no polariser for detection. In that case dye molecules at all azimuthal angles are evenly excited, there is no φ dependence in Eq. (4.7) and therefore Eq. (4.17) holds exactly for all n.

We can see that the sum of the experimental intensities I_{tot} depends on an even power of a cosine of the polar angle between the director and the xy plane. This angular dependence means that areas with director in the xy plane will be bright but there will be a strong intensity drop when the director is perpendicular to the xy plane. Total FCPM intensity I_{tot} will therefore give us information about the angle θ, but because the power is even, the sign of the angle is lost and $\theta \in [0, \pi/2]$ (we do not know if director is pointing up or down from the xy plane). Another way to understand this is that because all the probing polarisations lie in the xy plane, no information is gathered about the out-of-plane component of the director except its size, deduced form the size of the in-plane component and the fact that director is a unit vector. Therefore we can only calculate the director from the four FCPM intensities I_k up to the sign of the z-component: $\mathbf{n} = (\cos\varphi \cos\theta, \sin\varphi \cos\theta, \pm\sin\theta)$.

This lack of information about the θ sign prevents full reconstruction of the LC structure directly form experiment, but in some cases the structure can still be deduced from the available information. One of the options is to reconstruct the director "by hand" if the structure is not too complex i.e. it has some level of symmetry [5, 7, 12, 15]. In this case, you identify regions with high fluorescence intensity at each experimental polarisation and presume that the director in those areas is parallel to the polarisation. In favourable situations you can gather enough information about the director to deduce it in the remaining areas by interpolation. Another option is to compare the cross-sections of the experimental scans with theoretically predicted structures or even simulated fluorescence intensities calculated from the predicted structures to verify the similarity of the predicted and observed structure [16, 17]. If it is hard to identify which slices should be compared between experimental and theoretical data, it is sometimes useful to construct Pontryagin-Thom surfaces to visualise the data in 3D [14, 18].

One option for measurement of the sign of the out-of-image-plane director component is to use light with polarisation lying out of the xy plane. A trick to accomplish this, is to tilt the sample by some angle with respect to the optical axis of the microscope [19]. Because this tilt breaks the symmetry between positive and negative θ, the θ dependence in Eq. (4.17) is modified and the sign can be extracted. In this way it is possible to differentiate between director structures which would have identical intensities for non-tilted FCPM. There are a few hindrances that need to be considered when using this method, namely: (i) By tilting the sample, the non-index-matched boundary between the cover glass and the LC is not oblique to the optical axis of the microscope. This can shift the image and lower the resolution of the microscope. (ii) To achieve high optical resolution, oil immersion objectives need to be used. These objectives usually have short working distances of around 100 microns, which severely limits the tilt angle to a few degrees. As such low tilt angles, the contrast calculated from Eq. (4.17) can be comparable to the noise levels. (iii) The contrast gained by tilting the sample depends on the angle between φ and the direction of the polarisation/tilt. To extract information about the θ sign in all regions, the sample would need to be tilted along each of the polarisations. In complex director structures the number of measurements for each structure would have to be doubled, causing twice as much bleaching and extending the experiments. In our case the sample would have to be carefully tilted four times for each droplet which could induce movements in the droplets. The tilted data are also hard to correlate to zero-tilt data because of image shifting and reduced resolution because of the tilts. Taking these points into account, we find this method impractical for our experimental system and instead try to find an alternative path to find the θ sign, which is presented in Chap. 6.

So far we have discussed only the intrinsic limitations of the method but we haven't dealt with the shortcomings of its experimental implementation [7, 12, 13]. The fluorescence intensities of the FCPM measurements can be affected by all the maladies of fluorescent confocal microscopy due to refractive index mismatch such as spherical aberration and resulting reduction of resolution but because we are conducting FCPM in an optically non-uniform birefringent medium additional phenomena can affect the quality of images, e.g. depolarisation and polarisation guiding as discussed in Sect. 2.6.1. Fluorescence intensity is reduced with depth because of scattering and absorption. Intensity variation because of dye concentration gradients around defects is possible in principle, but so far it has been observed only with specific dyes [20]. A liquid crystal specific effect, which can affect FCPM is reorientation of LC molecules because of the electric field of the focused light. High excitation intensity can also heat the sample and affect the elastic constants, solubility and even the phase of the LC. Because of this, care should be taken during the experiments to keep the excitation intensity low enough not to interfere with the studied physical phenomena. We will discuss the other effects in detail in Chap. 6.

References

1. R. Oldenbourg, G. Mei, New polarized light microscope with precision universal compensator. J. Microsc. **180**, 140–147 (1995)
2. Y.-K. Kim, S.V. Shiyanovskii, O.D. Lavrentovich, Morphogenesis of defects and tactoids during isotropic-nematic phase transition in self-assembled lyotropic chromonic liquid crystals. J. Phys.: Condens. Matter **25**, 404202 (2013)
3. R.H. Webb, Confocal optical microscopy. Rep. Prog. Phys. **59**, 427–471 (1996)
4. M. Born, E. Wolf, *Principles of Optics*, 7th edn. (Cambridge University Press, 1999)
5. I.I. Smalyukh, S. Shiyanovskii, O. Lavrentovich, Three-dimensional imaging of orientational order by fluorescence confocal polarizing microscopy. Chem. Phys. Lett. **336**, 88–96 (2001)
6. B.J. Luther, G.H. Springer, D.A. Higgins, Templated droplets and ordered arrays in polymer-dispersed liquid-crystal films. Chem. Mater. **13**, 2281–2287 (2001)
7. I. Smalyukh, O. Lavrentovich, Three-dimensional director structures of defects in Grandjean-Cano wedges of cholesteric liquid crystals studied by fluorescence confocal polarizing microscopy. Phys. Rev. E **66**, 051703 (2002)
8. J. Rietdorf, E.H. Stelzer, Special optical elements, in *Handbook of Biological Confocal Microscopy* (2006), pp. 43–58
9. I. Smalyukh, R. Pratibha, N. Madhusudana, O. Lavrentovich, Selective imaging of 3D director fields and study of defects in biaxial smectic a liquid crystals. Eur. Phys. J. E **16**, 179–191 (2005)
10. G.H. Heilmeier, L. Zanoni, Guest-host interactions in nematic liquid crystals. A new electro-optic effect. Appl. Phys. Lett. **13**, 91–92 (1968)
11. R.J. Cox, Liquid crystal guest-host systems. Mol. Cryst. Liq. Cryst. **55**, 1–32 (1979)
12. S. Shiyanovskii, I. Smalyukh, O. Lavrentovich, Computer simulations and fluorescence confocal polarizing microscopy of structures in cholesteric liquid crystals, in *Defects in Liquid Crystals: Computer Simulations, Theory and Experiments* (Springer, 2001), pp. 229–270
13. R.P. Trivedi, T. Lee, K.A. Bertness, I.I. Smalyukh, Three dimensional optical manipulation and structural imaging of soft materials by use of laser tweezers and multimodal nonlinear microscopy. Opt. Express **18**, 27658–27669 (2010)
14. B.G.-G. Chen, P.J. Ackerman, G.P. Alexander, R.D. Kamien, I.I. Smalyukh, Generating the hopf fibration experimentally in nematic liquid crystals. Phys. Rev. Lett. **110**, 237801 (2013)
15. B. Senyuk et al., Topological colloids. Nature **493**, 200–205 (2012)
16. I.I. Smalyukh, Y. Lansac, N.A. Clark, R.P. Trivedi, Three-dimensional structure and multistable optical switching of triple-twisted particle-like excitations in anisotropic fluids. Nat. Mater. **9**, 139–145 (2010)
17. P.J. Ackerman, Z. Qi, I.I. Smalyukh, Optical generation of crystalline, quasicrystalline, and arbitrary arrays of torons in confined cholesteric liquid crystals for patterning of optical vortices in laser beams. Phys. Rev. E **86**, 021703 (2012)
18. B.G.-g. Chen, Topological defects in nematic and smectic liquid crystals. Ph.D. thesis, University of Pennsylvania, USA, 2012
19. O.P. Pishnyak, Y.A. Nastishin, O. Lavrentovich, Comment on "Self-organized periodic photonic structure in a nonchiral liquid crystal". Phys. Rev. Lett. **93**, 109401 (2004)
20. T. Ohzono, K. Katoh, J.-I. Fukuda, Fluorescence microscopy reveals molecular localisation at line defects in nematic liquid crystals. Sci. Rep. **6**, 36477 (2016)

Chapter 5
Materials and Experimental Setup

In this chapter we present the materials and procedures we use to optimise our experiments and analysis of the experimental data.

5.1 Materials

Dispersions of LC droplets are easily prepared by mixing small amounts of liquid crystals with an immiscible carrier medium. In our case, the selection of the two materials is dictated by their refractive indices to enable 3D sectioning of the sample by confocal microscopy, which is substantially hindered by mismatched refractive indices and birefringence.

5.1.1 Liquid Crystal Mixture

One of the most important steps to enable reconstruction of director fields is to obtain a liquid crystal with a birefringence as close to zero as possible. Typical values of liquid crystals used for optical applications are in the range 0.1–0.3, with some rare exceptions in the range 0.06–0.1. The liquid crystal mixture most often used for FCPM in the previous studies, ZLI-2806, has a birefringence of 0.045 [1], but is unfortunately difficult to obtain as some of its components are not produced commercially any more.

A group of liquid crystals with relatively low birefringence are $4\alpha,4'\alpha$-dialkyl-$1\alpha,1'\alpha$-bicyclohexyl-4β-carbonitriles or CCNs for short [2–5]. Figure 5.1 shows the structures of compounds 4'-butyl-4-heptyl-bicyclohexyl-4-carbonitrile (CCN-47) and 4,4'-dipentyl-bicyclohexyl-4-carbonitrile (CCN-55). CCN-47 was used in the original FCPM studies for observations of smectic structures [6, 7], but at

© Springer Nature Switzerland AG 2018
G. Posnjak, *Topological Formations in Chiral Nematic Droplets*,
Springer Theses, https://doi.org/10.1007/978-3-319-98261-8_5

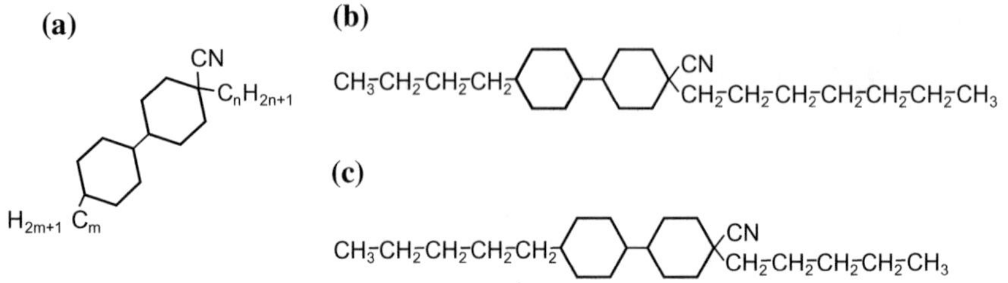

(a)

(b)

(c)

Fig. 5.1 Structures of $4\alpha,4'\alpha$-dialkyl-$1\alpha,1'\alpha$-bicyclohexyl-4β-carbonitriles **a** CCN-mn, **b** CCN-47, **c** CCN-55

Table 5.1 Phase sequences and refractive indices of CCN-47 and CCN-55. The refractive indices were measured in the nematic phase at 35 °C with an Abbe refractometer for light with a wavelength $\lambda = 589$ nm

LC	Phase sequence	n_o	n_e	Δn
CCN-47	SmA $\xrightarrow{30\,°C}$ N $\xrightarrow{60\,°C}$ I	1.469	1.500	0.031
CCN-55	SmB $\xrightarrow{30\,°C}$ N $\xrightarrow{65\,°C}$ I	1.469	1.502	0.033

temperatures above room temperature both compounds have a relatively wide nematic phase. Some basic physical properties of CCN-47 and CCN-55 are presented in Table 5.1 and examples of optical textures under crossed polarisers are shown in Fig. 5.2.

Experiments on chiral nematic droplets of CCN-47 need to be performed at an elevated temperature. If we heat the sample above the smectic-nematic transition temperature, temperature gradients are present between the heater stage and the microscope objective which is in contact with the sample through the immersion oil. The heat current induces a significant rotation of the droplet structure during the imaging process because of the Lehmann effect [8–10]. This rotation hinders the acquisition of full 3D scans of droplets at several polarisations which takes around 10 min. We tried to remove the temperature gradient by heating also the cover glass, but it proved difficult to stabilise the rotation of the droplet on the time scale of the measurement.

We solved the Lehmann rotation issue by finding a room-temperature low-birefringence nematic. Prof. Surajit Dhara, who has worked extensively with the different CCN components, suggested that mixing two different CCN compounds will lower the nematic-to-smectic transition temperature below the room temperature. We mixed CCN-47 and CCN-55 in a 1:1 weight ratio, and achieved a homogeneous nematic mixture with the nematic-to-isotropic (N-I) transition around 60 °C. The mixture is nematic at room temperatures but phase-separates around 15 °C. The mixture has the same refractive indices as its constituents.

The two chiral dopants we use to induce chirality in the CCN mixture are the right-handed (S)-4'-(2-Methylbutyl)[1,1'-biphenyl]-4-carbonitrile (CB15) and left-

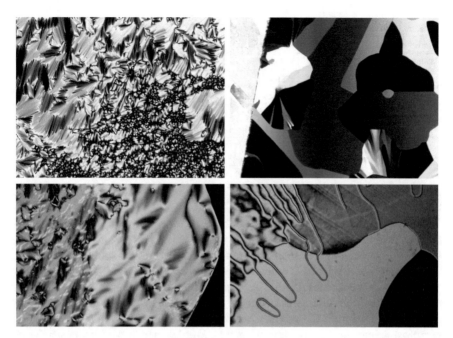

Fig. 5.2 Textures of CCN-47 and CCN-55 under crossed polarisers. Left column: CCN-47 in smectic A phase at 25 °C (top) and in the nematic phase at 40 °C (bottom). Right column: CCN-55 in smectic B phase at 25 °C (top; a part of a birefringent mylar spacer is visible on the left edge of the picture) and in the nematic phase at 59 °C (bottom). The samples were prepared with untreated glass

handed (S)-Octan-2-yl4-((4-(hexyloxy)benzoyl)oxy)benzoate (S-811). We measure their helical twisting power (HTP) from the positions of Grandjean-Cano disclination lines in a wedge cell with planar anchoring [11]. Pitch is measured for mixtures with several different concentrations and HTP is calculated from Eq. (2.3). We find that the two dopants have similar HTP in the CCN mixture as they do in 5CB: the average value for HTP at room temperature for CB15 is $6.3 \, \mu m^{-1}$ and for S-811 $8.7 \, \mu m^{-1}$.

5.1.2 Emulsion Matrix

After finding a LC with low birefringence, a suitable carrying medium has to be identified. The two main conditions for the medium are: it needs to induce homeotropic anchoring on the droplets and it must have a refractive index (RI) close to the LC mixture. Table 5.2 shows refractive indices of some possible media for LC emulsions together with the type of anchoring they induce. The medium with a RI closest to the LC mixture is glycerol, but on its own it induces planar anchoring. Fortunately,

Table 5.2 Possible media for LC droplet emulsions

Medium	Anchoring	Refractive index
Water	Planar	1.33
Glycerol	Planar	1.47
PDMS	Homeotropic	1.42
Water with 5 mM SDS	Homeotropic	1.33
Glycerol with 1% wt. lecithin	Homeotropic	1.47

surfactants CTAB and lecithin can be dissolved in it to change the anchoring to homeotropic [12–14].

Lecithin concentrations around 1% by weight in glycerol are sufficient for homeotropic anchoring. Higher concentrations also give homeotropic anchoring, but also significantly increase the viscosity of the mixture. We found that a lecithin concentration of 4% had sufficient viscosity to prevent movement of the droplets, so we used this concentration for the experiments. In specific mixtures of LC the glycerol/lecithin mixture can change the type of anchoring with temperature [14] but in the case of the CCN mixture no such effects were observed at elevated temperatures neither in the structure of nematic droplets nor in the light transmission of a homeotropic cell under crossed polarisers, with the glasses coated with the glycerol/lecithin mixture.

The lecithin used in our experiments is isolated from egg yolk and its proper name is L-α-phosphatidylcholine. Chemically it is a glycerol molecule with the – OH groups substituted with long carbohydrate chains. The L in its name signifies it is a left-handed chiral molecule which leads to an unexpected consequence: if a right-handed dopant CB15 is used for the chiral mixture, the droplets exhibit a significant reduction in chirality. We attribute this effect to an interaction of the left-handed lecithin and right-handed chiral dopant which somehow reduces the effective chirality, but we do not understand the exact mechanism. With a left-handed chiral dopant S-811 there are no observable effects of lecithin on the chirality of droplets.

5.1.3   Fluorescent Dye

To enable FCPM, the LC mixture has to be doped with a fluorescent dye. Following the guidance of the original FCPM studies [6, 7], we used a dye *N,N'*-bis(2,5-di-*tert*-butylphenyl)-3,4,9,10-perylenedicarboximide, in FCPM literature called BTBP, but also known as DBPI, with a structure shown in Fig. 5.3.

The dye itself has poor solubility in the CCN mixture. If it is mixed directly into the liquid crystal, large crystals of undissolved dye are visible under a microscope and the solution has a dull colour, suggesting that the dye is not sufficiently dissolved to fluoresce. Heating the LC mixture to isotropic phase helps the dissolution, but the

Fig. 5.3 Structure of the dye
N,N′-bis(2,5-di-*tert*-
butylphenyl)-3,4,9,10-
perylenedicarboximide
(BTBP or DBPI) [15, 16]

Fig. 5.4 Spectrum of BTBP
absorption and emission in
the CCN mixture. Emission
was measured for a sample
excited with wavelengths in
the band 480–490 nm. The
excitation wavelengths used
in our experiments are
shaded in red, and the
detection band in yellow

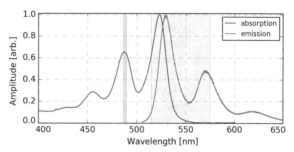

colour stays dull. Much better dissolution can be achieved if the dye is first dissolved in a solvent, for example ethanol. This brightly coloured solution of the dye in ethanol can be added to a LC. In our case, mixing a suitable amount of dye/ethanol solution with the CCN mixture results in an isotropic mixture, which is left at room temperature overnight for the alcohol to evaporate. The next day, the mixture is nematic and brightly fluorescent. To remove any residual alcohol, the mixture is heated for a couple of hours to 90 °C. We check for the presence of impurities by measuring the N-I phase transition of this mixture and it shows no variation compared to the mixture without the dye. The absorption and emission spectrum of BTBP dye, dissolved in the CCN mixture, is shown in Fig. 5.4.

5.1.4 Sample Preparation

The droplet dispersion is prepared by placing a small volume of glycerol with lecithin, roughly corresponding to the cell volume, on a thin microscopy cover glass. Liquid crystal is added to glycerol by dipping a sharp needle in the LC mixture so that a small amount of LC was on the tip of the needle, then placing the tip of the needle in glycerol and mixing the glycerol with circular moves of the needle. The number of moves determines the size distribution of LC droplets, the average size of droplets deceasing with continued mixing. Usually 5–10 gyrations are enough to obtain a dispersion in which most of the droplets are in the 10–20 μm diameter range, which is the most suitable for FCPM microscopy. The number of droplets is controlled with the amount of LC inserted into glycerol with the needle. In samples which are the most suitable for microscopy, the number of droplets is low enough that a typical droplet is not in contact with or obscured by other droplets.

After mixing the LC and glycerol, mylar spacers of desired thickness are placed next to the glycerol droplet and the mixture is covered with a glass plate of approximate dimensions $12 \times 12 \times 1$ mm. The LC mixture has lower density than the glycerol/lecithin matrix, so the LC droplets migrate towards the top glass of the cell because of their buoyancy. Because resolution of microscopy deteriorates with sample thickness, the spacers are not much thicker than the droplet diameter—typically 30μm. After the glycerol has had time to spread throughout the cell, the top glass is pressed down and any excess glycerol coming out on the sides is wiped away. The cell is then sealed with a two-component epoxi glue around the circumference to suppress glycerol flow in the cell.

Quenches of the droplets are conducted by placing the sealed sample cell on a hot plate set to a temperature above the N-I phase transition temperature of the LC mixture, typically to $70 \,^\circ$C. After the sample is heated to the isotropic phase, it is removed from the hot plate and placed on a surface at room temperature to cool down. The rate of cooling can be increased by blowing compressed nitrogen over the sample.

5.2   Microscopy Setup

We use two types of microscopes for this study: a typical upright microscope shown in Fig. 4.1, which we use for polarised images, and a confocal microscope.

The confocal microscope is a Leica TCS SP5 X with the white light laser (WLL) shown in Fig. 5.5a. The system is based on a motorised inverted microscope Leica DMI6000B. The light source for excitation is a laser which generates a continuous visible spectrum of wavelengths 470–670 nm with a photonic-crystal fiber

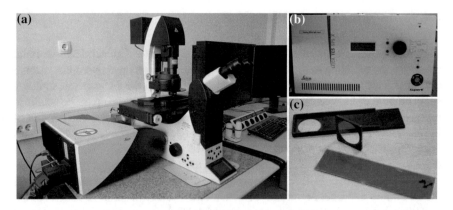

Fig. 5.5 The confocal microscopy setup. **a** Leica TCS SP5 X confocal unit is attached to the left side port of the Leica DMI6000B inverted microscope. **b** The Leica White Light Laser (WLL). **c** A custom-made frame (black plate) for insertion of the $\lambda/4$ waveplate (centre) and a linear polariser (bottom) in the polariser slot of the microscope

(Fig. 5.5b). The light from the source is passed through an acousto-optic transmission filter (AOTF) which selects a narrow band of wavelengths from the continuous spectrum, typically 1–2 nm wide. Up to 8 bands of excitation light can be selected at the same time by the AOTF. The output of the AOTF is coupled to the main optical path by an acousto-optic beam splitter (AOBS) which deflects the excitation light so it is parallel to the main optical path. The excitation beam is deflected by a pair of galvo mirrors, moving in perpendicular directions to preform the xy scanning. The beam enters the microscope from a side port and is focused on the sample by an oil immersion objective (Leica HCX PL APO lambda blue 63.0 × 1.40 OIL UV).

Even though the output of the WLL light source is unpolarised due to the photonic-crystal fiber, the excitation light at the objective is linearly polarised, because the effect of both AOTF and AOBS is polarisation dependent. To utilise FCPM to its full extent, we need to control the polarisation of the excitation light. We accomplish this by inserting a $\lambda/4$ retardation plate for the excitation wavelength 488 nm in to the beam path at a 45° angle to the polarisation of light coming from the AOBS. This transforms the polarisation of excitation light from linear to circular. Circular polarisation itself is useful for FCPM if an even excitation of all dye molecules lying in the xy plane is desired to measure the out-of-plane tilt of the director. A linear polarisation of excitation light is achieved by placing a linear polariser between the $\lambda/4$ waveplate and the objective. Half of the excitation intensity is lost because of the linear polariser, but this can be compensated for by increasing the laser power. Four polarisers sandwiched between glass plates were prepared with suitable orientations, so that the 4 different linear polarisations needed in the FCPM reconstruction procedure could be selected by exchanging the polarisers. Both the $\lambda/4$ waveplate and the linear polariser are placed in a small frame shown in Fig. 5.5c, which fits in the polariser slot of the microscope below the objective.

The fluorescence is collected through the objective, passing through the linear polariser, resulting in polarised detection. The subsequent $\lambda/4$ waveplate changes the polarisation of the fluorescent light, but as no part of the fluorescence light path after the polariser is polarisation sensitive, this has no influence on the detected intensity. The fluorescent light leaves the microscope through the same side port as the excitation light came in. In the confocal unit, it is de-scanned by the galvo mirrors and passes through the AOBS without deflection. Next, the beam is passed through the confocal pinhole, which filters out most of the fluorescence intensity coming from out-of-focus planes. The fluorescent light is then dispersed into its spectrum by a prism and an adjustable slit determines which wavelengths are detected by a photomultiplier.

As described in Sect. 4.2, to reconstruct the director field from FCPM intensities, imaging is done at four polarisations of excitation and detection separated by $\pi/4$ (polarisations angles 0, $\pi/4$, $\pi/2$ and $3\pi/4$ relative to x axis in the xy plane, thus gaining intensities I_0, I_1, I_2 and I_3). The imaging at each polarisation is done by scanning the laser beam in the xy plane over the whole field of view to obtain a single horizontal slice at a given depth z and then changing the focus by one step in the z direction and repeating the xy scan to image the next slice. In this way we obtain 3D stacks of xy images for each of the probing polarisations. By doing this we discretise the sample. Each data point in the 3D stack corresponds to a small volume in the sample—a voxel. After the 3D scans are performed for each polarisation, the scan at the first polarisation is repeated to evaluate the bleaching rate from the reduction of intensity. The bleaching rate is used to correct the intensities I_0, I_1, I_2 and I_3 in the data analysis stage as described in Sect. 6.2.1.

The scan of the xy plane is performed by galvo mirrors which move the focused excitation beam line by line. The typical pixellation of the xy plane is 512×512, and the typical line scanning rate 2800 Hz, so about 5 scans of the xy plane can be performed in a second at these settings. Usually, 3 or 4 consecutive scans of the same plane are averaged to enhance the signal-to-noise ratio. Scanning in the z direction is done with a stepper motor which moves the objective to change the focus. The parameters of the scan are selected so that the steps in the z direction are larger than the xy size of the pixels by a factor of 3. Here we need to take into account that the difference in refractive indices between the sample and immersion oil effectively stretches the z dimension—if the objective is moved by d, the focus in the sample is changed by $d\,n_{\mathrm{medium}}/n_{\mathrm{immersion}}$. Physical size of the 3D voxel was typically 40–67 nm for the xy and 120–200 nm for the z dimension. These voxel dimensions are considerably smaller than the details which can still be resolved by an optical microscope, as we have discussed in Chap. 4. This oversampling is, as we will see later, required to perform deconvolution [17, 18]. Because of the physical size of the droplets, around 150 z-slices have to be collected for a full 3D scan at a single polarisation, which takes 2–3 min. The intensity data is digitised with a 12 bit dynamic range.

5.2.1 Numerical Processing of Data and Visualisation

Fluorescence intensities of pixels in each xy slice are recorded as uncompressed tiff images. A series of tiff images for each 3D scan at a given polarisation is imported into MATLAB, where the intensity data are numerically processed for the reconstruction procedure. The only part of the reconstruction procedure performed outside MAT-LAB is deconvolution of the FCPM images. The deconvolution is done on the 3D stacks of FCPM images for each polarisation in a commercial program SVI Huygens Professional [19]. The simulated annealing algorithm is written as a MATLAB function in C++ to optimise the speed. For a typical droplet with around 10^6 voxels, the simulated annealing algorithm described in Chap. 6 takes around 5 min on a single core of an Intel i7 processor. The annealed director field is saved as a 3D vector field along with additional scalar fields such as experimental intensities in a single Visualization Toolkit (VTK) file [20].

The VTK files are imported into ParaView [21] for 3D manipulation and visualisation. Director fields are visualised with cylindrical rods which show the director orientation. An image of the structure can be generated as a snapshot in ParaView or alternatively in POV-Ray [22] rendering software which uses ray tracing to simulate illumination effects and gives more control over the appearance of the image.

The visualisation of director fields with streamlines was done by projecting the director in each pixel in a plane to that plane and using the orientation of the projection to draw a line of a given length. Each such line was assigned a random grey-scale colour to randomise the pattern of the streamlines. This procedure is a variant of line integral convolution [23] but because we are projecting a 3D vector field to a 2D surface, one of the components of the director field is lost in the procedure. In areas where the director is perpendicular to the surface of projection, the size of the projected director is small and therefore the streamlines do not contain much information. Because of this we do not draw the streamlines in areas where the projection of the director is smaller than $1/3$. In this way we turn the 3D vectors in discrete points into a continuous 2D map from which it is easy to recognise the topological features of the director field, as we will see in the Chap. 7. A single such a map does not reveal the direction of helical rotation, so in some cases it is more useful to draw the director field with cylinders. The code for generating the streamlines is implemented in C and used in this Thesis by the courtesy of Simon Čopar. The images with streamlines are rendered in POV-Ray.

References

1. I. Smalyukh, O. Lavrentovich, Three-dimensional director structures of defects in Grandjean-Cano wedges of cholesteric liquid crystals studied by fluorescence confocal polarizing microscopy. Phys. Rev. E **66**, 051703 (2002)
2. B.S. Scheuble, G. Weber, R. Eidenschink, Liquid crystaline cyclohexylcarbonitriles: properties of single compounds and mixtures. Proc. Eurodisplay **65** (1984)
3. S. Dhara, N. Madhusudana, Enhancement of the orientational order parameter of nematic liquid crystals in thin cells. Eur. Phys. J. E **13**, 401–408 (2004)
4. S. Dhara, N. Madhusudana, Influence of director fluctuations on the electric-field phase diagrams of nematic liquid crystals. Europhys. Lett. **67**, 411 (2004)
5. S. Dhara, N. Madhusudana, Physical characterisation of 4′-butyl-4-heptyl-bicyclohexyl-4-carbonitrile. Phase Transitions **81**, 561–569 (2008)
6. I.I. Smalyukh, S. Shiyanovskii, O. Lavrentovich, Three-dimensional imaging of orientational order by fluorescence confocal polarizing microscopy. Chem. Phys. Lett. **336**, 88–96 (2001)
7. S. Garg, K. Purdy, E. Bramley, I. Smalyukh, O. Lavrentovich, Electric field-induced nucleation and growth of focal-conic and stripe domains in a smectic a liquid crystal. Liq. Cryst. **30**, 1377–1390 (2003)
8. O. Lehmann, Structur, system und magnetisches verhalten flüssiger krystalle und deren mischbarkeit mit festen. Annalen der Physik **307**, 649–705 (1900)
9. J. Yoshioka et al., Director/barycentric rotation in cholesteric droplets under temperature gradient. Soft Matter (2014)
10. G. Poy, P. Oswald, Do Lehmann cholesteric droplets subjected to a temperature gradient rotate as rigid bodies? Soft Matter (2016)
11. P. Pieranski, Classroom experiments with chiral liquid crystals, in *Chirality in Liquid Crystals* (Springer, 2001), pp. 28–66
12. S. Candau, P. Le Roy, F. Debeauvais, Magnetic field effects in nematic and cholesteric droplets suspended in a isotropic liquid. Mol. Cryst. Liq. Cryst. **23**, 283–297 (1973)
13. M. Kurik, O. Lavrentovich, Negative-positive monopole transitions in cholesteric liquid crystals. J. Exp. Theor. Phys. Lett. **35**, 444–447 (1982)

14. G. Volovik, O. Lavrentovich, Topological dynamics of defects: boojums in nematic drops. Zh. Eksp. Teor. Fiz. **85**, 1997–2010 (1983)
15. W. Ford, H. Hiratsuka, P. Kamat, Photochemistry of 3,4,9,10-perylenetetracarboxylic dianhydride dyes. 4. spectroscopic and redox properties of oxidized and reduced forms of the bis(2,5-di-tert-butylphenyl)imide derivative. J. Phys. Chem. **93**, 6692–6696 (1989)
16. S. El-Daly, M. Okamoto, S. Hirayama, Fluorescence quenching of N, N′-bis(2,5-di-tert-butylphenyl)-3,4:9,10-perylenebis(dicarboximide) (dbpi) by molecular oxygen. J. Photochem. Photobiol. A: Chem. **91**, 105–110 (1995)
17. J.B. Pawley, Points, pixels, and gray levels: digitizing image data, in *Handbook of Biological Confocal Microscopy* (Handbook of Biological Confocal Microscopy, 2006), pp. 59–79
18. V. Centonze, J.B. Pawley, Tutorial on practical confocal microscopy and use of the confocal test specimen, in *Handbook of Biological Confocal Microscopy* (Handbook of Biological Confocal Microscopy, 2006), pp. 627–649
19. SVI Huygens Professional. https://svi.nl/HuygensProfessional
20. Kitware Visualisation Toolkit. http://www.vtk.org/
21. ParaView. http://www.paraview.org/
22. POV-Ray. http://www.povray.org/
23. B. Cabral, L.C. Leedom, Imaging vector fields using line integral convolution, in*Proceedings of the 20th Annual Conference on Computer Graphics and Interactive Techniques* (1993), pp. 263–270

Chapter 6
Simulated Annealing for Determination of z-Component Sign

In this chapter we first present the simulated annealing algorithm with which we augment the existing FCPM method to enable full reconstruction of director fields from experimental FCPM data. We test the algorithm on idealised FCPM data, calculated from complex, numerically obtained structures in chiral nematic droplets and characterise its success rate. Then we discuss the important parameters when implementing the algorithm on experimental data and how to process the experimental data to optimise the reconstruction procedure. Finally, we demonstrate the full director reconstruction procedure on three different structures: a nematic droplet, a chiral nematic droplet with a complex director structure and a toron in a chiral nematic cell with homeotropic anchoring.

As shown previously, the director can be determined from FCPM intensities through Eqs. (4.14) and (4.17) only up to the sign of n_z: $\mathbf{n} = (\cos\varphi\cos\theta, \sin\varphi\cos\theta, \pm\sin\theta)$. The two possible orientations for each point are reminiscent of a spin-$1/2$ system, which has two possible states for each spin—either up or down. The state of a whole system of such spins can be found by using a simulating annealing (SA) algorithm, which was originally developed as an optimisation algorithm that can be applied to various problems, e.g. the travelling salesman problem or optimisation of connection lengths in an electronic chip [1]. In the case of spins, this algorithm changes the state of each spin with a probability dependent on the difference of interaction energies of the two possible states of a spin with its neighbours and a temperature parameter, which is slowly lowered to find a stable configuration. The physical interaction between spins is different from the elastic interactions in the LC, but because the complexity of the problem is similar, the same algorithm can be successfully implemented.

During the FCPM experiment, the sample is divided into 3D voxels because of the raster nature of our data acquisition. Director orientation $(\varphi, \pm\theta)$ in each cube voxel is calculated according to Eqs. (4.10), (4.14) and (4.17) from the FCPM intensities at four polarisations, separated by $\pi/4$. A voxel is randomly selected by the algorithm and an evaluating energy E is calculated for the starting state of the director and for

© Springer Nature Switzerland AG 2018
G. Posnjak, *Topological Formations in Chiral Nematic Droplets*,
Springer Theses, https://doi.org/10.1007/978-3-319-98261-8_6

a state with an inverted sign of n_z. Later on we will test two different forms of the evaluating energy E, so we will not specify its exact form at this point. Here we only need to know that E describes the interaction of the director in a given voxel with its neighbouring voxels, and will therefore change during the run of the algorithm as the director field in the neighbouring voxels will have different configurations.

The algorithm keeps the inverted sign of n_z in the chosen location if its evaluating energy is lower than the starting one, but if it is higher, the sign is kept with a probability dependent on a Boltzmann factor

$$P = \exp(-\Delta E/t) , \qquad (6.1)$$

where ΔE is the difference between the evaluating energies of the original and the inverted sign of θ in that voxel, and t is a free parameter which acts as an effective temperature. The calculation is repeated at a given t for random voxels until the evaluating energy of the whole sample does not stabilise and then t is lowered. The procedure starts at a high t, effectively randomising the signs of n_z in all voxels and runs until lowering the effective temperature does not change the total energy of the sample any more. The acceptance of unfavourable configurations with a probability dependent on the effective temperature helps the algorithm not to become stuck in local minima and thus to scan the configuration space much more efficiently.

A special consideration is needed for the voxels on the surface of the droplet. In principle the anchoring contribution to the total free energy of the sample would need to be calculated for these points [2], but in our case the exact anchoring energy is not very important because we only need to choose from two available orientations of the director in each point on the surface. Because we are studying droplets with homeotropic anchoring, the orientation of director on the surface is known a priori—it has to be normal to the surface. Therefore we can omit the expression for the anchoring energy and substitute it with a simpler method of determining the correct orientation. We manage this by adding a fixed radial director field outside the droplet (Fig. 6.1). By doing so, we can treat all the points in the droplet in the same way, because minimisation of our evaluating energy ensures there is the least amount of deformation in the director. Therefore the fixed radial configuration outside the droplet promotes radial orientation of director on the inside surface, effectively mimicking homeotropic anchoring.

6.1 Testing the Algorithm

To test the annealing algorithm, we used director structures of chiral nematic droplets with homeotropic anchoring provided by David Seč. These structures were obtained in numerical quenches which start with random orientation of director in each point of the droplet and then relax the structure by minimising Landau-de Gennes elastic free energy of the droplet [2]. Results of these numerical experiments were presented in Ref. [3]. Because of the frustration between the homeotropic anchoring on the surface

(a)   **(b)**   **(c)**

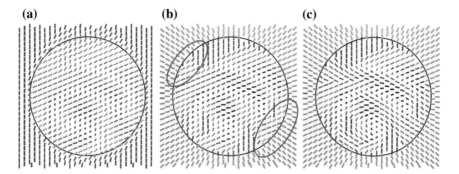

Fig. 6.1 Anchoring correction of the director field in a xz plane. The boundary of the droplet is marked with a black line. **a** Director field calculated from the raw intensity data. The calculated director outside the droplet, where there is no liquid crystal or dye, is vertical because the total fluorescence intensity I_{tot} is low. The director orientation in the outside volume has no physical meaning, because the liquid outside the LC droplet is isotropic. **b** The calculated director outside the droplet is substituted with a director field, which points towards the centre of the droplet (green cylinders). We can see that the director marked with magenta ellipses does not match the anchoring conditions. **c** The annealed director in the droplet matches the outside radial director everywhere on the surface

of the droplets and the twisting of the chiral nematic liquid crystal, cholesteric layers cannot occupy the whole volume and consequently many topological defects appear in places, where the director is frustrated. The defects were found to take the shape of extended line defects close to the surface of the droplet or even to run through the bulk of the droplet. In some cases the defect lines can form knots and links. Because of such diversity of states, we considered the structures sufficiently complex to verify the effectiveness of operation of the simulated annealing algorithm.

We used the numerically obtained structures to calculate simulated FCPM intensities by projecting the director to the four polarisations of excitation and detection according to Eq. (4.8), with n and a both equal to 1, corresponding to our experimental method of single photon excitation and polarised detection. With this we gained data on which we could test and calibrate the simulated annealing algorithm and verify its accuracy by comparing the annealed structure to the original, numerically obtained director. We used a correlation parameter C to test the accuracy of the algorithm, calculated as the average of the absolute value of the scalar product between the annealed and original director in all points in the droplet:

$$C = \frac{1}{M} \sum_{i=1}^{M} |\mathbf{n}_{i,\text{orig}} \cdot \mathbf{n}_{i,\text{anneal}}| , \qquad (6.2)$$

where M is the number of points in the droplet and i is the index that runs over them. The value of the scalar product is 1 if the original and annealed director are the same in a given point and -1 if they are anti-parallel. Because of the nematic symmetry $\mathbf{n} = -\mathbf{n}$, anti-parallel vectors are equal, therefore we have to consider only the absolute value of the scalar product. We will see from examples in the next chapter that the correlation C is sensitive enough to enable validation of successful reconstruction of the director.

6.1.1 Continuity as the Evaluating Criterion
in the Boltzmann Factor

First, we tested the algorithm by taking the ΔE in Eq. (6.1) to be the sum of the absolute value of scalar products of the director in a given point with its neighbours:

$$\Delta E_{i,\text{scalar}} = \sum_{j} |\mathbf{n}_i \cdot \mathbf{n}_j| , \qquad (6.3)$$

where i is a point in the droplet chosen randomly during the annealing and j runs over its closest neighbours in a cubic grid. With the choice of scalar product for the energy function, the annealing algorithm effectively searches for a director which is continuous everywhere.

We ran the algorithm with $\Delta E_{i,\text{scalar}}$ 100 times on each numerical droplet to characterise its performance. The statistics of these runs are presented in Fig. 6.2. We can see, that the SA algorithm with Eq. (6.3) for an energy function is not deterministic as the annealed director structures have a wide range of correlations with the original numerical director and that the typical correlation of the annealed structure is around 90%. Three different examples of cross-sections of reconstructed director fields for one of the numerical droplets are shown in Fig. 6.3—the lowest, average and the highest C out of the 100 runs (Fig. 6.3a, b and c, respectively).

Such an algorithm finds a continuous director almost everywhere, but there are almost always some discontinuities present (marked with cyan arrows in Fig. 6.3). More precisely, the algorithm successfully finds the correct sign of z-component almost everywhere in the droplet (Fig. 6.3c), but in many cases there are domains with a wrong choice of the sign in the annealed structure. This is because in areas where the director is either completely horizontal or vertical (green arrows in Fig. 6.3), both signs will result in a continuous director but the wrong choice of sign will propagate from that region through the structure and result in a domain of misaligned director. Such a domain will end on the other side with a discontinuity of the director— a domain wall. The domain walls do move around during the SA procedure, but disappear only if the domains shrink in volume by moving the wall to a region of horizontal director. Because of this, such domain walls are relatively stable and difficult to remove from the structure.

We can see that the algorithm is very efficient at ensuring the director field is continuous, but fails because of formation of domains with wrong orientation. These domains grow from areas of horizontal director where the z sign can be changed continuously. This suggests that the best way to remove domains of wrong z-component

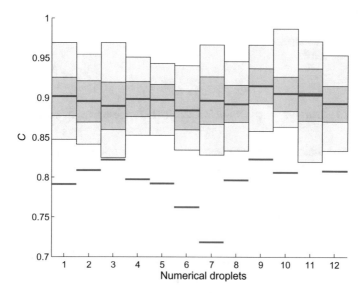

Fig. 6.2 Correlation factor C after the annealing for different numerical droplets. Red lines show the correlation C before the annealing, blue the average finishing value of C for each droplet in 100 runs of the SA algorithm. Dark gray areas show the standard deviation around the mean value and light gray areas the whole range of the finishing C values in the 100 runs of the SA algorithm for each droplet

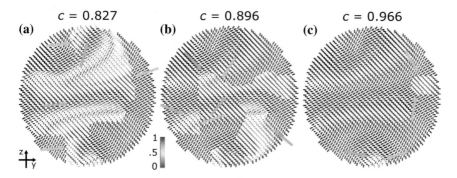

Fig. 6.3 Examples of the annealed director in a vertical cross-section for one of the numerical droplets. The cylinders show the annealed director and their color the local correlation between the annealed and original director of the numerical droplets. Red areas have perfect correlation and blue are perpendicular to the original orientation. Panels **a** and **c** show the annealed director with the highest and lowest correlation in 100 runs of the SA algorithm and **b** shows a typical result of SA for a run with C close to the average value. The arrows mark sharp (cyan) and continuous (green) borders of domains

sign is to ensure that the right sign is chosen in these continuous borders. As we will see in the next section, the easiest way to do this in chiral nematics is to include a chiral energy term.

6.1.2 Boltzmann Factor with Landau-de Gennes Free Energy

Next, we test the algorithm by using the one constant nematic Landau-de Gennes elastic free energy [Eq. (2.11)] with the chiral term [Eq. (2.13)] for ΔE in the Boltzmann factor [Eq. (6.1)]. The order parameter Q at each point is calculated from the director field through Eq. (2.1), where the scalar order parameter S is taken to be constant.

The spatial derivatives in the equations are approximated by a mixture of symmetric and asymmetric finite differences [4]. The mixing of both types of derivatives ensures better numerical stability of the annealing algorithm. All the terms that change when the sign of n_z is inverted are included in the calculation. In practice this means we need to calculate also the energies of all the neighbours of the processed point and therefore the directors of up to two spaces away are included in the calculation. This gives our calculation a "persistence length" of 5 units which smooths the noise in the data and enables the domains of aligned director to grow.

First we run the algorithm scanning over a wide range of q values to examine the effect of chirality on the SA algorithm. Figure 6.4 presents data for one of the numerical droplets (droplet no. 1). For large negative values of q, the annealing is not successful—the annealed director has a worse matching to the original one than the reconstructed director before annealing. For small values of q the algorithm is quite successful regardless of the sign of q (effectively neglecting the chiral term), finding

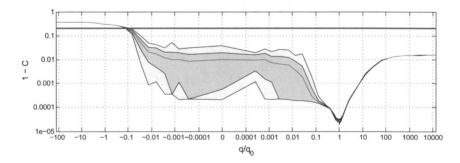

Fig. 6.4 The deviation $1 - C$ of the annealed director from the original one, for different values of q in Eq. 2.13. The red line shows the correlation C before the annealing, the blue the average finishing value of C in 15 runs of the SA algorithm for each q, dark gray area shows the standard deviation around the mean value and light gray area the whole range of the finishing C values in the 15 runs for each q. The q on the x axis are normalised with q_0, the pitch with the best matching between the annealed structure and the original director

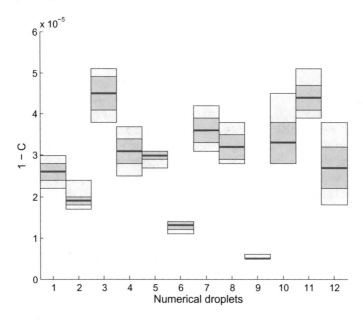

Fig. 6.5 Deviation $1 - C$ of the annealed director from the original one for different numerical droplets. Blue lines show the average finishing value of deviation $1 - C$ for each droplet in 20 runs of the SA algorithm, dark gray area the standard deviation around the mean value and light gray area the whole range of the finishing $1 - C$ values in the 20 runs for each droplet

an annealed director which on average matches the original one within 1 percent. This is an order of magnitude better performance than the algorithm with the scalar product from the previous section, meaning that Landau-de Gennes free energy is a much better tool for discerning between the two possible director orientations even without chirality. This is because the Landau-de Gennes free energy minimises director distortions in addition to ensuring continuity of the director, which is the only thing the scalar product does. For larger positive values of q the algorithm performs even better, with a wide range of q-values matching the original director to better than 0.01%. Therefore the chosen value of q is not very important for the success of the SA algorithm, as long as the right chirality is used.

We performed a series of algorithm runs with full chiral elastic energy at the optimal value of q also for the other numerical droplets. We can see in Fig. 6.5 that the average annealed director matches the original one within 0.01%, so the performance gain compared with just using a director continuity criterion as in the previous section is three orders of magnitude, regardless of the complexity of the structure in the droplet. This clearly illustrates the suitability and stability of the simulated algorithm for determination of the z-component sign which is missing in the experimental data.

6.2 Implementing the Simulated Annealing Algorithm on Experimental Data

In the previous section, we have demonstrated the operation of the simulated anneal-ing algorithm on idealised FCPM data, which was calculated from director fields obtained in numerical quenches. In experimental data, there are many factors which can hinder the annealing process: experimental noise, bleaching, blurring of the image because of the finite size of the point spread function of the optical system, background signal in areas where director is perpendicular to the probing polarisa-tion, polarisation guiding and attenuation of intensity with depth. To ensure success-ful reconstruction of director fields we have to take these phenomena into account or at least try to mitigate their effects as much as possible.

There are some measures we can take to minimise these effects which reduce the quality of experimental data. The first and perhaps the most important one is to choose a suitable experimental system. To ensure high resolution of confocal microscopy images, spherical abberation must be prevented by choosing an index matched experimental system. With the materials presented in the previous chapter, the miss-match is relatively low: the refractive index of the glass we used for the cells was 1.52, the index of glycerol 1.47 and the refractive indices of the CCN LC mixture $n_o = 1.47$ and $n_e = 1.50$. This index matching minimises the blurring of images because of birefringence of the LC and reduces spherical aberrations.

Another important factor is the pitch of the chiral nematic, which can in conjunc-tion with birefringence cause polarisation guiding. By calculating the conditions discussed in Sect. 2.6.1, we can see from:

$$0.5 \, p \, n \, \Delta n = 0.16 \, \mu\text{m} \ll \lambda \ll p = 7 \, \mu\text{m} \, , \tag{6.4}$$

we are in the short wavelength circular regime, in which the plane of polarisation rotates with a rate calculated from Eq. (2.32): $\rho = -\pi (\Delta n)^2 p/(4\lambda^2) \approx 1°/\mu\text{m}$. This means that at depths of about $20 \, \mu\text{m}$ in the LC, the plane of polarisation is rotated by about $20°$, and can lead to misinterpretation of intensity data at different polarisations. Fortunately, this effect is pronounced only if light is travelling along the chiral axis of the structure, which is rarely the case in droplets. If the light is travelling along the layers, the birefringent sample acts as an optical lens because of the spatially varying refractive index, locally focusing or dispersing the light. The defocusing due to the birefringence can be estimated as $z\Delta n$ where Δn is the birefringence of the LC and z is the depth of scanning [5, 6]. In our samples this defocusing is smaller than $0.5 \, \mu\text{m}$ except in the largest droplets.

The next important step is to choose a suitable laser power for FCPM. If the laser power is too high, it causes significant photo-bleaching of dye molecules and can heat up the sample because of absorption and distort the director field through the Fréedericksz effect due to the electric field of the focused laser light [7, 8]. With low laser powers, the signal to noise ratio of detected intensities is low and the data are not suitable for reconstruction. A way to increase the signal to noise ratio is to

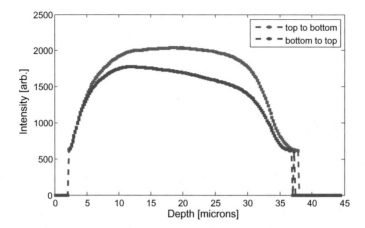

Fig. 6.6 Dependence of fluorescence intensity on the z scanning direction. Average fluorescence intensity in each z slice is shown. If the scan is started in the xy plane closest to the objective (starting from $z = 0 \mu$m; blue line), the intensity will fall with the depth of scanning because of absorption and also with time because of bleaching. In this scenario both effects reduce the fluorescence intensity with increasing depth. In our case the magnitude of both effects is approximately equal, so if we reverse the direction of z scanning, the two effects almost cancel out (red line)

collect several frames at a given depth and average them. In this way we can control the influence of laser beam on the structure, but the measurements are prolonged and bleaching is still present because dye molecules go through more excitation cycles [9].

An important experimental step is to choose a suitable concentration of the fluorescent dye. The concentration should be high enough to give sufficient signal, but not as high as to strongly attenuate the excitation beam intensity with depth or disturb the ordering of the LC. Influence of the dye on the nematic phase can be easily tested by measuring the N-I phase transition, as the transition temperature is lowered if impurities interfere with nematic order [10]. Attenuation of the excitation intensity with depth can be easily measured by performing a 3D FCPM scan of the droplet in the isotropic phase. In this way we remove the influences of the birefringent structure and scattering and only attenuation and bleaching remain. Because attenuation increases with depth and photo-bleaching with exposure, we can partly mitigate these effects by choosing a suitable direction of scanning in the z direction. If we start the scanning with the layers which are furthest away from the objective, the intensities will be lowered by absorption, but not by bleaching. By the time the scanning reaches the layers closest to the objective, intensities will be lowered because of bleaching but there will be less absorption. Dye concentration can be adjusted so that these two effects at least partially cancel each other out, ensuring a flatter intensity depth profile during a single 3D scan as seen in Fig. 6.6.

Even with these experimental measures the FCPM data are still not suitable for quantitative reconstruction of director fields because of bleaching between 3D scans at different polarisations, experimental noise and low contrast, background signal and attenuation of intensity with depth. The next few subsections present post-acquisition procedures that handle these effects to prepare the data for application of the SA algorithm.

6.2.1 Bleaching Corrections

There are two main causes for reduction of fluorescence intensity during a prolonged exposure: excitation to long-lived triplet states and photo damage [11, 12]. In each excitation cycle of a dye molecule there is a small probability that the molecule will not be excited to short-lived singlet fluorescent state with a typical lifetime on the time scale of nanoseconds, but instead to a triplet state which does not have an allowed transition to the ground state and therefore relaxes only slowly, making it long-lived (on the time scale of micro- to millisecond) [11]. Because this state does not fluoresce during the dwell time of the sensor on that sample volume, it suppresses fluorescence. Photo damage of the dye molecules is a less understood effect, for which several possible mechanisms have been proposed. The basic effect is that under illumination the fluorophore reacts with the surrounding chemicals, which renders it irreversibly non-fluorescent. In practice this means that with increasing illumination dosage the proportion of dye molecules which fluoresce is decreasing. Because our droplets are small, isolated volumes of LC doped with dye, this effect can be very pronounced, because the bleached dye molecules are not replaced with fresh ones by diffusion.

We already discussed how bleaching during a 3D scan at a single polarisation can be partially mitigated by choosing a suitable direction of scanning and dye concentration. Acquisition of the 3D scans at all polarisations typically takes 10 min. Such long exposures can cause significant bleaching which needs to be corrected for to enable quantitative analysis of data. In our experiments bleaching causes the intensities to drop by 10–50%, depending on the droplet size, dye concentration and laser intensity. To correct the effects of bleaching between the 3D scans, we repeat the imaging of the droplet at the polarisation 0 after all the other polarisations were taken. In this way we can calculate how much the integral intensity of the droplet, calculated as the sum of the intensities of all points in the droplet, has reduced from the first 3D scan to the last one. We then use this data to linearly interpolate the bleaching rate for the 3D scan at each polarisation and correct the intensities by the appropriate factor.

6.2.2 Deconvolution

As we have seen in Sect. 4, a point object is not imaged as a point through a microscope, but instead has a finite sized point spread function, which is a property of the imaging system and of the wavelength of the used light. Because of this, the fine

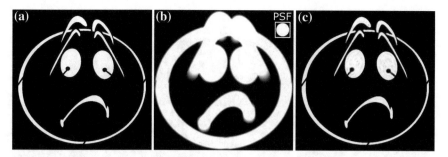

Fig. 6.7 Example of an image (**a**), blurred by a point spread function so that fine details are obscured (**b**), and then deconvolved with Matlab (**c**). The function `deconvlucy` was used with the original blurring PSF and 100 iterations. Note that the dynamical range of the convolved (t.i. experimental) image has to be large enough to ensure the intensity values are added linearly

details in the original object appear bigger in the image and in some cases cannot be resolved. We can illustrate this effect by taking a grayscale image (Fig. 6.7a) and generate a new image in which each point is replaced with a bigger point (Fig. 6.7b). Such an image will show less details than the original one, but because the grayscale values linearly add in areas where the PSFs overlap, information about the original distribution of intensity is still available. This is true in the case of incoherent light for which the amplitudes of intensities add instead of the amplitudes of electric field and the final image can be considered a linear sum of the individual PSFs of the objects, without interference effects. Such a process is mathematically called convolution [13]. In this case, it is possible to reconstruct the original object through an inverse process called deconvolution, if we know the PSF of the imaging system (Fig. 6.7c).

During deconvolution, the contributions of the neighbouring pixels to a given pixel are subtracted in an iterative procedure which "sharpens" the image a bit with each consecutive step. Figure 6.7 illustrates that with proper knowledge of the PSF even fine details such as the thin gaps between white areas, which were obscured in Fig. 6.7b are clearly visible after deconvolution. In practical situations deconvolution is hindered by imperfect knowledge of the PSF and by experimental noise, but with proper implementation, the procedure is still able to improve microscopic images in many situations. Many different procedures have been developed for deconvolution, but we won't present any in detail as we use a commercial product SVI Huygens Professional, which is dedicated to restoration of microscopy images.

An additional effect of deconvolution is the reduction of noise in the experimental image. When subtracting the influence of surrounding voxels, the software takes values of neighbouring voxels into account to generate a better estimate of the intensity in a specific voxel [14], thus improving the signal to noise ratio.

An important factor for successful deconvolution is the quality of the experimental data. Ideally, the system should be RI matched and the optics should be ideally compensated so that there is as little distortion to PSF as possible. Unfortunately, complete RI matching is not possible in our case, because of birefringence on the one

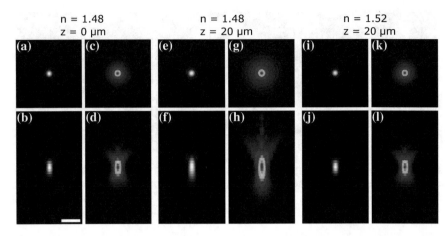

Fig. 6.8 Examples of numerically calculated PSFs for different refractive indices of medium at different depths in the sample. The top row shows the PSF in the xy plane and the bottom row in the xz plane. For each combination of refractive index and depth the left column shows the intensity in grayscale and the right the maximum intensity projection (MIP) of the intensity in pseudocoloured scale with gamma correction set to 1.7 to highlight the dimmer regions. All PSF are calculated for an oil immersion objective with $NA = 1.4$ and refractive index of the oil 1.518. The left and center PSF correspond to our experimental system at depths 0 and 20 μm. We can see the PSF at $z = 20$ μm (**e–h**) is substantially enlarged compared to the one in a perfectly RI matched sample, shown in (**i–l**). The pinhole size is set to one Airy unit. The PSF were calculated using the SVI calculator available at: https://svi.nl/NyquistCalculator (accessed 22. 3. 2017). The scale bar is 1 μm

hand and of the limited choices of LC and the medium on the other. The algorithm in Huygens can handle RI mismatch as we will see later, but not birefringence. Despite this, deconvolution was able to improve the quality of our data as we will see from comparing raw and deconvolved data. Furthermore, the data should be sampled close to the Nyquist sampling rate which guaranties that all the spatial frequencies that are needed for reconstruction at maximum resolution are present in the experimental data [9, 15]. In our case, where we used excitation light with wavelength 488 nm and an objective with 1.4 numerical aperture, the ideal sample size is 43 nm [16], the typical sample size is 40 nm but sample sizes up to 67 nm were used successfully.

We used a theoretical model for the PSF of our system, which was calculated based on the parameters of our experiment—the refractive indices of the immersion oil and the medium, numerical aperture of the objective, confocal pinhole size and the wavelength of the fluorescence light. One of the effects the Huygens Professional software is capable of taking into account, is the spherical aberration because of the mismatch between the immersion and medium refractive index. Because of the refractive index mismatch, the beams that pass the interface between the two mediums are refracted which causes them not to focus in an optimal PSF. The effect worsens with depth so the software uses different theoretical PSFs for different focusing depths (Fig. 6.8).

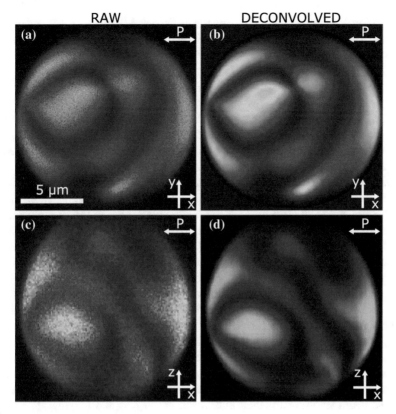

Fig. 6.9 Example of image quality improvement from raw (**a, c**) to deconvolved (**b, d**) images for a horizontal (**a, b**) and a vertical (**c, d**) cross-section of a chiral nematic droplet

An example of deconvolved FCPM images of a LC structure in a chiral nematic droplet with homeotropic anchoring is shown in Fig. 6.9. In Fig. 6.9b we can notice a substantial improvement of signal to noise ratio compared to Fig. 6.9a and in Fig. 6.9d a significant improvement of contrast, especially in areas between bright regions along the z direction can be seen.

After deconvolution the data is downsampled in the xy dimension, so that cube voxels are obtained, with a side length of 120–200 nm, depending on the size of the droplet.

6.2.3 Offset Correction and Normalisation

At this point the data is smooth enough to calculate the director field from Eqs. (4.10), (4.17), (4.14) and (4.4). However, such a director field would contain significant artefacts, because the fluorescent signal does not drop to zero when the probing polarisation is perpendicular to the director as expected from Eq. (4.17) and because intensity data needs to be normalised. These two effects have to be taken into account by modifying the Eq. (4.17):

$$I_{tot} = I_{offset} + I_{norm} \cos^{2n+2} \theta \, , \tag{6.5}$$

where I_{offset} is the value of the background signal which is present in areas where the director is vertical, I_{norm} is the normalisation set by I_{tot} in areas where the director lies in the xy plane and n is a constant, marking the order of the process used for excitation of the dye molecules. In our case of FCPM, $n = 1$.

The non-zero signal in areas where the director is perpendicular to the probing polarisation is a consequence of scattering, non-zero order parameter of the LC and the dye and of the distortion of the linear polarisation in the strongly focused excitation laser beam. For normalisation of data, it is not simply enough to take the highest intensity value and use it as normalisation, as the maximum intensity changes with depth because of absorption, scattering and ray optics effects such as spherical aberration and defocusing due to refractive index mismatch and birefringence. These effects are mitigated as much as possible with the choice of proper materials and the procedure described in previous sections, but they still cause significant reduction of FCPM signal as will be demonstrated on concrete examples later on. These effects do not affect the calculation of the in-plane angle φ, because the background subtracts in Eq. (4.14) and the normalisation factors cancel out, so only the polar angle θ is affected.

A way to deal with these effects is basically to fit the z dependence of minimum and maximum values of the total FCPM intensity I_{tot} with a function. For simple structures, a reasonable choice would be a linear function as we expect the effects at least to a first approximation to increase linearly with depth. For droplets with complex cholesteric structures we expect that there will be at least some area in each z-slice with vertical and horizontal director where I_{tot} should be minimal and maximal, respectively. Therefore, we should get a good approximation of proper background offset and normalisation values, if we plot the values of lowest and highest I_{tot} in each z layer.

This procedure is complicated by the unevenness of the effects and by the inherent noise in the fluorescent intensities. Therefore the z-dependent values of I_{offset} and I_{norm} are smoothed and additionally corrected by a factor to ensure wide regions of horizontal and vertical director, which enable smooth transitions between domains of differently oriented oblique director. Typically the offset values are increased by about 10% and the normalisation values are decreased by about 5%. Such intensity clipping enhances the stability of the SA algorithm on experimental data.

6.3 Calibration of the Annealing Algorithm on Experimental Data

In this section we demonstrate the implementation of the simulated annealing algorithm and the calibration of the different parameters on experimental data of three different director structures. The reconstruction of a nematic droplet with radial director structure was presented in Ref. [17] and that of a skyrmion in a flat homeotropic cell in Ref. [18].

6.3.1 Nematic Homeotropic Droplet

One of the simplest structures we can use for testing and calibrating our algorithm is a nematic droplet with homeotropic anchoring on its surface, in which the director forms a radial structure [17]. We prepared the droplets in the same way as the chiral nematic droplets, but using the CCN mixture without the chiral dopant.

Raw experimental FCPM intensities in several orthogonal cross-sections for this structure are shown in Fig. 6.10. Figure 6.10g shows a graph of integrated FCPM intensities of all the voxels inside of the droplet for each of the probing polarisations, from which the bleaching rates are calculated. We can see that in this case the intensity in the droplet falls by 20 % between the first and last scan. The bleaching-corrected integrated intensities in Fig. 6.10g are much closer to each other in value than the uncorrected ones. After the bleaching correction, the 3D stack for each of the polarisations is deconvolved with the program SVI Huygens Professional. A few examples of deconvolved images are shown in Fig. 6.11.

The next step after deconvolution is correcting the I_{tot} and running the SA algorithm. The variation of minimum and maximum I_{tot} in each z layer of the droplet is shown in Fig. 6.12a. If the algorithm is ran with $I_{offset} = 0$ and $I_{norm} = \max(I_{tot})$, the resulting director field is discontinuous in areas where it should be vertical or horizontal as can be seen in Fig. 6.12b. To obtain a smooth structure we have to therefore appropriately adjust I_{offset} and I_{norm} to obtain vertical and horizontal orientations of director in the reconstruction.

Because in a radial structure areas of horizontal and vertical director are not present in each z layer of the droplet, we cannot estimate I_{offset} and I_{norm} from the lowest and highest I_{tot} values at all depths in Fig. 6.12a, but instead have to adjust the two levels to obtain a suitably smooth structure without excessive artefacts. Because of the radial director structure, we expect the director at the top and bottom of the droplet to be mostly vertical, so we can use these areas to estimate I_{offset}. Likewise, the director should be mostly horizontal in an equatorial xy plane so we can base our estimate of I_{norm} on these regions.

We ran the SA algorithm with several different linear estimates of I_{offset} and I_{norm} shown in Fig. 6.12a. The resultant annealed director fields in a vertical plane through the centre of the droplet are shown Fig. 6.12c–e. If I_{offset} and I_{norm} values are not

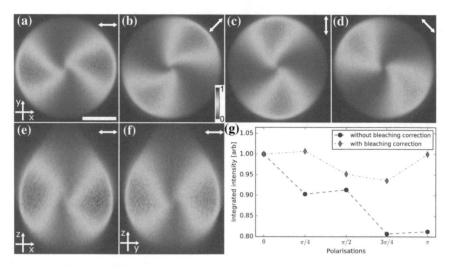

Fig. 6.10 Raw experimental FCPM data for a nematic droplet with a radial structure. Panels **a–d** show intensities for the 4 probing polarisations in an equatorial xy plane and **e, f** the intensities for two orthogonal vertical planes going through the center of the droplet. Panel **g** shows a graph of integrated intensities of the droplet for each of the 5 polarisations (the final polarisation at $0°$ is labelled π) before and after the bleaching correction. The order of acquisition was $0, \pi/2, \pi/4, 3\pi/4, \pi$. The scale bar is $5\,\mu$m

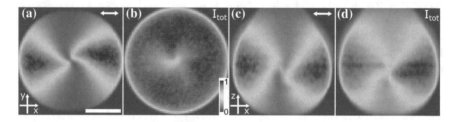

Fig. 6.11 Deconvolved FCPM data for a nematic droplet with a radial structure. Intensity of a single polarisation (**a, c**) and the sum of the 4 probing polarisations I_{tot} (**b, d**) in an equatorial xy plane (**a, b**) and in a vertical xz plane (**c, d**) going through the center of the droplet. The scale bar is $5\,\mu$m. Reprinted from Ref. [17]. This figure is distributed under the terms of the Creative Commons Attribution 4.0 International License (http://creativecommons.org/licenses/by/4.0/)

overestimated as described in the previous section, there is still some variation of normalised I_{tot} within each z layer because of geometrical and other factors—in the present case the intensity drops off towards the edge of the droplet. If we reconstruct the director from such data, we can see that there is a discontinuity at the outer parts of the droplet close to the equator because the director cannot make a smooth transition between the two signs of n_z. This is because the corrected I_{tot} are not large enough for the director to be horizontal and facilitate a continuous transition between the two orientations. Therefore we have to underestimate the I_{norm} values to compensate

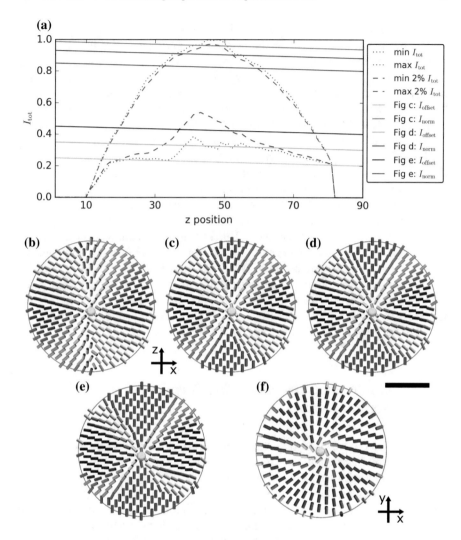

Fig. 6.12 Examples of SA algorithm runs on a radial nematic droplet with different settings for I_{offset} and I_{norm}. **a** A graph of I_{tot} in each xy layer. The blue lines show the minimal value (dotted line) and the average value of the lowest 2% (dashed line) of I_{tot} in the droplet for each z layer. The red lines show analogous maximal values and the solid lines show I_{offset} and I_{norm} settings for the different runs of the SA algorithm. **b–e** The director obtained with SA algorithm for different settings of I_{offset} and I_{norm}. In **b** $I_{\text{offset}} = 0$ and $I_{\text{norm}} = \max(I_{\text{tot}})$, for the rest the values are shown in (**a**). Notice the abrupt jumps in (**b, c**) and excessive artefacts in (**e**). **f** The reconstructed director in an equatorial xy plane. The yellow point marks the position of the point defect. The scale bar is 5 μm. Panels d and f reprinted from Ref. [17], distributed under the terms of the Creative Commons Attribution 4.0 International License (http://creativecommons.org/licenses/by/4.0/)

for these intensity variations and ensure wide regions of horizontal director to obtain smooth reconstructed director. Similar reasoning leads to increasing the I_{offset} value.

We can see that incorrect settings of the offset and normalisation result either in sharp jumps in the director fields (Fig. 6.12b, c) or in excessive artefacts due to clipping of the intensities (Fig. 6.12e), which cause large areas of vertical and horizontal director with only narrow transition regions of oblique director. The best result is obtained for an intermediate setting of I_{offset} and I_{norm} with some artefacts (wider vertical and horizontal regions than would be expected for a structure with minimised free energy) but still with wide transitional regions of oblique director. Figure 6.12d shows the reconstructed director at the optimal levels of I_{offset} and I_{norm}. We can see that the whole procedure nicely reconstructs the radial structure in the xz plane and even shows the characteristic twisting of the radial structure in the xy plane which minimises the elastic energy of the structure, due to the twist elastic constant being smaller than the splay constant [19]. The normalisation and offset settings strongly affect the out-of-plane angle, which means we can take it's value only as an estimation of the experimental tilt. We should note that even with these incorrect settings with excessive artefacts, the transition areas may become narrower or wider but they stay at the same location. This means that the topology of the director field is preserved which can also be seen from the point defect which is present in the centre of the structure at all of the presented settings.

6.3.2   Chiral Nematic Droplet with a Complex Director Structure

A similarly instructive example of the reconstruction procedure with a more systematic method of determining the offset and normalisation can be shown on a chiral nematic droplet with homeotropic anchoring at $N \approx 7$. Figure 6.13a–f shows raw FCPM intensities in a xy (Fig. 6.13a–d) and a xz (Fig. 6.13e, f) plane going through the centre of the droplet. From the variation of the fluorescence intensities we can see that the director changes its direction many times inside the droplet and the striped intensity pattern suggests a layer-like structure. Figure 6.13g shows the integrated intensities of the droplet for each polarisation before and after the bleaching correction. We can see that the fluorescence intensity reduced by 40% between the 0 and π polarisations, but with the repeated measurement of the starting polarisation, the rest of polarisations can be bleaching corrected. Large differences between the integrated intensities at different polarisations can be seen because of the anisotropy of the structure in the droplet.

Figure 6.14a, b show two deconvolved polarisations in an equatorial xy slice, Fig. 6.14c the sum of the four polarisations I_{tot} in the same slice and Fig. 6.14d I_{tot} in a vertical slice. We can see from Fig. 6.14d that each z layer of the droplet includes all the possible values of I_{tot}.

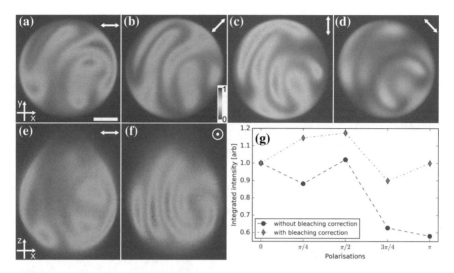

Fig. 6.13 Raw experimental FCPM data for a chiral nematic droplet with a complex director structure ($N \approx 7$). Panels **a–d** show intensities for the 4 probing polarisations in an equatorial xy plane and **e, f** intensities at two orthogonal polarisations in a vertical plane going through the centre of the droplet. Panel **g** shows a graph of integrated intensities of the droplet for each of the 5 polarisations (the final polarisation at $0°$ is labelled as π) before and after the bleaching correction. The order of acquisition was $0, \pi/2, \pi/4, 3\pi/4, \pi$ and the scale bar is $5\,\mu m$

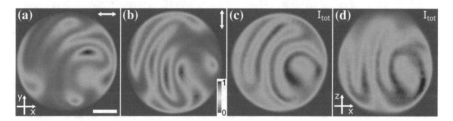

Fig. 6.14 Deconvolved FCPM data for the chiral nematic droplet from Fig. 6.13. **a, b** Intensity of a single polarisation and **c** the sum of the 4 probing polarisations I_{tot} in an equatorial xy plane and **d** I_{tot} in a vertical xz plane going through the centre of the droplet. The scale bar is $5\,\mu m$

Figure 6.15a shows the variation of the minimal and maximal values of I_{tot} with the z coordinate. The minimum I_{tot} which correspond to vertical orientations of director are in this case reasonable approximations of I_{offset} and maximum I_{tot} values correspond to horizontal director and can be used for I_{norm}. The yellow and the green lines in Fig. 6.15a show the estimates for I_{offset} and I_{norm} which we use for the reconstruction. These estimates are calculated from smoothed average values of the 2% of pixels with lowest and highest I_{tot} values in each layer and additionally increased by 12% and decreased by 6%, respectively. On the top and bottom of the droplet the intensity values are reduced because of the proximity of the edge of the droplets. In those parts we approximate the I_{offset} and I_{norm} by a linear function as

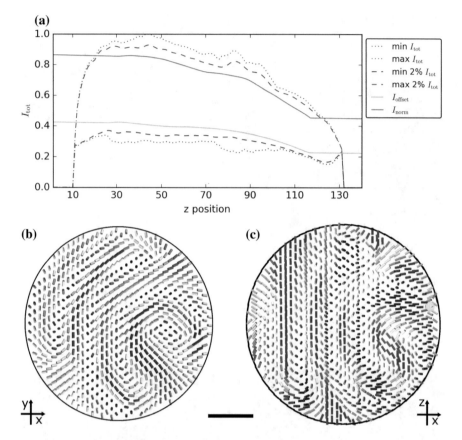

Fig. 6.15 Correction of I_{tot} and the reconstructed director field in a layered droplet. **a** Variation of minimal and maximal I_{tot} values with the depth in the droplet. The I_{offset} and I_{norm} values we used for the reconstruction are marked with yellow and green lines, respectively. **b, c** The reconstructed director field in a xy (**b**) and a xz plane (**c**). The point defect is marked with yellow. The scale bar is $5\,\mu m$

seen in Fig. 6.15a. By using these values for correction of I_{tot}, the SA algorithm obtains a smooth reconstructed director as shown in Fig. 6.15b, c.

The director field in Fig. 6.15b, c is shown with cylinders which are coloured by the I_{tot}. We can see that the striped areas of high I_{tot} really are layers of director field lying in the xy plane, which are separated by low-intensity areas where the director is twisted out of the xy plane. In Fig. 6.15b we can see that in this case the layers are bent in the xy plane and run approximately vertically as seen in Fig. 6.15c. The droplet has a single point defect marked with a yellow dot in Fig. 6.15c, which is consistent with the topological charge in a droplet with homeotropic anchoring.

6.3.3   Skyrmions and Torons in Homeotropic Cells

To illustrate that our procedure of reconstruction is useful also for other systems, we reconstruct the director field in the well-known structure of a localised cholesteric deformation in a bubble domain texture, which is also known as a toron [20–25]. The structure has been discussed theoretically [22] and a numerically calculated structure was compared with FCPM intensities [23, 24]. Here we show how the structure of a toron can be reconstructed directly from experimental FCPM intensities [18].

The sample is prepared by assembling a 6–10 μm thick wedge cell with glasses treated with dimethyloctadecyl[3-(trimethoxysilyl)propyl]ammonium chloride also known as DMOAP for homeotropic anchoring and filling the cell with the CCN mixture with 1.2% S-811 ($p = 9.6$ μm). Individual torons spontaneously form at certain rations of pitch to cell thickness, but they do form more easily if the cell is

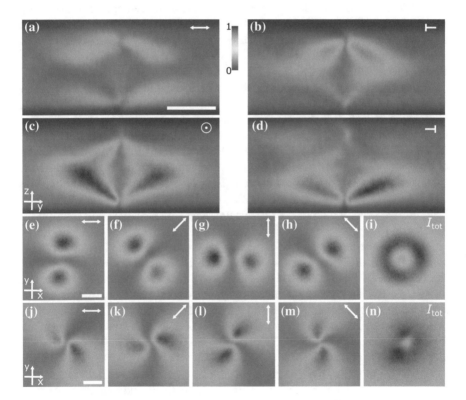

Fig. 6.16 Raw experimental FCPM data for a toron structure. Panels **a**–**d** show intensities for the 4 probing polarisations in a vertical yz plane going through the centre of the structure. **e**–**i** Intensities in a xy plane in the middle of the cell. **j**–**n** Intensities in a xy plane going through the centre of the bottom point defect. The scale bars in **a** and **e** are 5 μm, and in **j** 2 μm

perturbed, for example with a quench or with alternating electric field. We perform a temperature quench after which the torons form in parts of the cell with thickness around 8 μm.

Figure 6.16a–d show the individual fluorescence intensities in a vertical yz plane, going through the centre of the toron. We can see from the intensities, that the toron has a central region where all the intensities are very low, indicating a vertical director orientation. This region is surrounded by a region which is brightest in the out-of-plane polarisation, indicating that the director is perpendicular to the plane of the cross-section. The other intensities show that closer to the top and bottom of the structure, the director slightly twists towards the plane of the cross-section. Figure 6.16e–i show intensities in a xy plane going through the middle of the structure. We can see that in the region around the central part, the director is oriented in the azimuthal direction and that the structure is cylindrically symmetric. Figure 6.16j–n show fluorescence intensities in a xy plane going through the bottom point defect. The intensity patterns of individual polarisations are rotated compared to the ones in Fig. 6.16e–i, which agrees with our conclusion that the director twists towards the top and bottom parts of the structure. The central low intensity region in Fig. 6.16n is much smaller than in Fig. 6.16i, indicating that the central region of vertical director shrinks towards the point defect.

Figure 6.17a shows the sum of the four experimental intensities I_{tot} in the same vertical plane as Fig. 6.16a–d. We can see that the highest intensities are slowly dropping with depth, which is also confirmed by the plot of the maximum intensities in each layer (red lines) shown in Fig. 6.17b. In this structure there are pixels with lowest and the highest intensities at each z level, therefore we can base our estimates of I_{offset} and I_{norm} on the minimal and maximal values of I_{tot}. We approximate the average of the intensities with lowest and highest 2% values with two linear functions and offset both functions as shown in Fig. 6.17b with the yellow and green lines. With this we achieve clipping of the dynamic range of I_{tot} which will stabilise the SA algorithm.

The reconstructed director fields, which were calculated with these settings are shown in Fig. 6.18. Figure 6.18a shows the director field in the same vertical yz plane as in Figs. 6.16c–f and 6.17a. The reconstructed director is smooth everywhere and we can see the left-handed rotation through the central part of the structure. The structure appears discontinuous around the point defects, but from a close-up of the bottom defect in Fig. 6.18b we can see, that this is an effect of the averaging of the director structure. In fact this inset clearly shows a hyperbolic profile of the director around the point defect.

Figure 6.18c shows the director field in a xy plane in the middle of the cell. We can see that the director really is oriented azimuthally and that the structure is cylindrically symmetric. In this cross-section, the toron has a skyrmionic profile. Figure 6.18d shows the director around the bottom point defect in a xy plane. We can see that the director close to the defect is not completely radial as in a normal hyperbolic defect but somewhat twisted towards an azimuthal orientation.

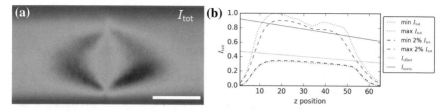

Fig. 6.17 Deconvolved FCPM data for the toron structure and minimal and maximal values of I_{tot} at each depth. **a** I_{tot} in a vertical xz plane going through the centre of the structure. **b** The graph of the I_{tot} values together with I_{offset} and I_{norm} used in the reconstruction. The scale bar is $5\,\mu m$

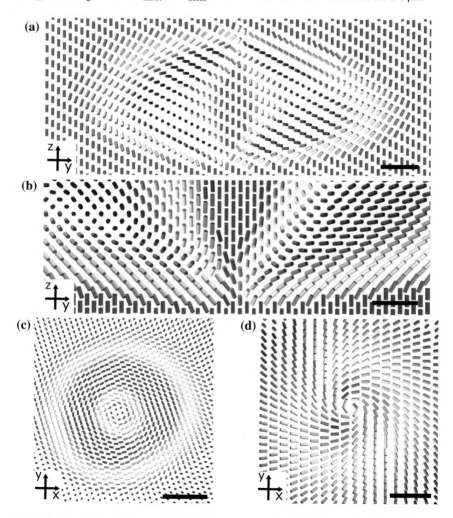

Fig. 6.18 Reconstructed director field for the toron structure. **a** The director field in a vertical yz plane. **b** A close-up of the director around the bottom point defect. **c** The director in a xy plane in the middle of the cell. **d** The director in a xy plane around the bottom point defect. The scale bars are **a** $2\,\mu m$, **b** $1\,\mu m$, **c** $5\,\mu m$ and **d** $2\,\mu m$

References

1. S. Kirkpatrick, C. Gelatt Jr., M. Vecchi, Optimization by simulated annealing. Science **220**, 671–680 (1983)
2. M. Ravnik, G.P. Alexander, J.M. Yeomans, S. Žumer, Mesoscopic modelling of colloids in chiral nematics. Farad. Discuss. **144**, 159–169 (2010)
3. D. Seč, S. Čopar, S. Žumer, Topological zoo of free-standing knots in confined chiral nematic fluids. Nat. Commun. **5**, 3057 (2014)
4. W.H. Press, S.A. Teukolsky, W.T. Vetterling, B.P. Flannery, *Numerical Recipes in C*, Vol. 2 (Cambridge University Press, 1982)
5. I. Smalyukh, O. Lavrentovich, Three-dimensional director structures of defects in Grandjean-Cano wedges of cholesteric liquid crystals studied by fluorescence confocal polarizing microscopy. Phys. Rev. E **66**, 051703 (2002)
6. S. Shiyanovskii, I. Smalyukh, O. Lavrentovich, Computer simulations and fluorescence confocal polarizing microscopy of structures in cholesteric liquid crystals, in *Defects in Liquid Crystals: Computer Simulations, Theory and Experiments* (Springer, 2001), pp. 229–270
7. P.G. de Gennes, J. Prost, *The Physics of Liquid Crystals*, 2 edn. (Clarendon Press, Oxford, 1993)
8. M. Škarabot et al., Laser trapping of low refractive index colloids in a nematic liquid crystal. Phys. Rev. E **73**, 021705 (2006)
9. V. Centonze, J.B. Pawley, Tutorial on practical confocal microscopy and use of the confocal test specimen, in *Handbook of Biological Confocal Microscopy* (Handbook of Biological Confocal Microscopy, 2006), pp. 627–649
10. S. Singh, Phase transitions in liquid crystals. Phys. Rep. **324**, 107–269 (2000)
11. R.Y. Tsien, L. Ernst, A. Waggoner, Fluorophores for confocal microscopy: photophysics and photochemistry, in *Handbook of Biological Confocal Microscopy* (2006), pp. 338–352
12. A. Diaspro, G. Chirico, C. Usai, P. Ramoino, J. Dobrucki, Photobleaching, in *Handbook of Biological Confocal Microscopy* (Handbook of Biological Confocal Microscopy, 2006), pp. 690–702
13. M.B. Cannell, A. McMorland, C. Soeller, Image enhancement by deconvolution, in *Handbook of Biological Confocal Microscopy* (2006), pp. 488–500
14. J.B. Pawley, Fundamental limits in confocal microscopy, in *Handbook of Biological Confocal Microscopy* (Handbook of Biological Confocal Microscopy, 2006), pp. 20–42
15. J.B. Pawley, Points, pixels, and gray levels: digitizing image data, in *Handbook of Biological Confocal Microscopy* (Handbook of Biological Confocal Microscopy, 2006), pp. 59–79
16. SVI Huygens Professional. https://svi.nl/HuygensProfessional
17. G. Posnjak, S. Čopar, I. Muševič, Points, skyrmions and torons in chiral nematic droplets. Sci. Rep. **6**, 26361 (2016)
18. A. Varanytsia et al., Topology-commanded optical properties of bistable electric-field-induced torons in cholesteric bubble domains. Sci. Rep. **7**, 16149 (2017)
19. H. Stark, Physics of colloidal dispersions in nematic liquid crystals. Phys. Rep. **351**, 387–474 (2001)
20. W.E. Haas, J.E. Adams, New optical storage mode in liquid crystals. Appl. Phys. Lett. **25**, 535–537 (1974)
21. M. Kawachi, O. Kogure, Y. Kato, Bubble domain texture of a liquid crystal. Jpn. J. Appl. Phys. **13**, 1457 (1974)
22. S. Pirkl, P. Ribiere, P. Oswald, Forming process and stability of bubble domains in dielectrically positive cholesteric liquid crystals. Liq. Cryst. **13**, 413–425 (1993)
23. I.I. Smalyukh, Y. Lansac, N.A. Clark, R.P. Trivedi, Three-dimensional structure and multistable optical switching of triple-twisted particle-like excitations in anisotropic fluids. Nat. Mater. **9**, 139–145 (2010)

24. P.J. Ackerman, Z. Qi, I.I. Smalyukh, Optical generation of crystalline, quasicrystalline, and arbitrary arrays of torons in confined cholesteric liquid crystals for patterning of optical vortices in laser beams. Phys. Rev. E **86**, 021703 (2012)
25. J. Gilli, S. Thiberge, D. Manaila-Maximean, New aspect of the voltage/confinement ratio phase diagram for a confined homeotropic cholesteric. Mol. Cryst. Liq. Cryst. **417**, 207–213 (2004)

Chapter 7
Structures in Chiral Nematic Droplets with Homeotropic Anchoring

In this chapter we present and analyse some of the structures we find in chiral nematic droplets with homeotropic anchoring. The structures can be roughly divided into those with layered cholesteric structures and those with several point defects separated by toron-like cholesteric structures, which we call cholesteric bubbles. First we present the simplest droplets that appear at low chiralities and include a single point defect and one cholesteric bubble. Then we move on to the typical layered structures with one or several point defects between the cholesteric layers. We show that surface line defects can also appear in these droplets. Next, we present droplets in which multiple point defects are separated by cholesteric bubbles and can also include newly discovered higher-charge topological point defects. At the end of the chapter we introduce topological molecules, where single point defects are substituted by bigger structures carrying equivalent topological charge.

Examples of typical droplets we observe in our samples are shown in Fig. 7.1. We can see that the smallest droplets are relatively simple with different variants of a dark cross with bent arms shown in the polarised images. The bigger droplets mostly have layer-like features, similar to the ones of the droplet, which we used as an example of the reconstruction procedure in Sect. 6.3.2. Some of the droplets show more specific textures, which we present in detail later in this chapter. First we take a look at the smallest droplets with the simplest structure.

7.1 Droplet with a Single Point Defect and a Cholesteric Bubble

Droplets with the lowest chirality ($N < 2$) have a radial structure with a point defect in the centre of the droplet. Some twisting is possible around the point defect, because splay deformation is energetically more costly than twist, just like in nematic droplets.

© Springer Nature Switzerland AG 2018
G. Posnjak, *Topological Formations in Chiral Nematic Droplets*,
Springer Theses, https://doi.org/10.1007/978-3-319-98261-8_7

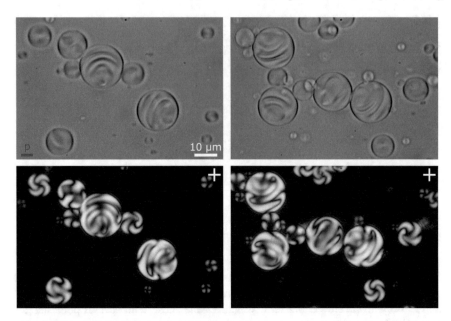

Fig. 7.1 Typical droplets which appear in our samples. The images in the top row are unpolarised and in the bottom row are images of the same droplets under crossed polarisers, oriented as indicated in the top right corners. The length of pitch of the CCN mixture with 2% of S-811 chiral dopant is shown with the red bar

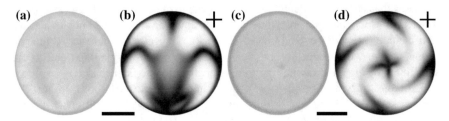

Fig. 7.2 Droplets with a single expelled point defect at different orientations. In **a, b** the symmetry axis of the droplet is perpendicular to the optical axis of the microscope and in **c, d** it is aligned with it. Images **b** and **d** are taken under crossed polarisers oriented as indicated with the cross in the top right corners of images. In **b** the point defect is positioned where the two "petals" in the bottom of the image meet, and in **d** it is at the central dark cross. Scale bars are 5 μm. Panel **a** and **c** reprinted from Ref. [1] under the terms of the Creative Commons Attribution 4.0 International License (http://creativecommons.org/licenses/by/4.0/)

With increasing $N \approx 2-4$, the first effect of the chirality is that the $+1$ point defect, which is necessary because of topological constraints, is expelled from the bulk towards the surface of the droplet as can be seen in transmission images in Fig. 7.2. In Fig. 7.2a we can see the point defect in the bottom part of the image as the point where the two dark brushes meet at the surface or the droplet. In Fig. 7.2c the point

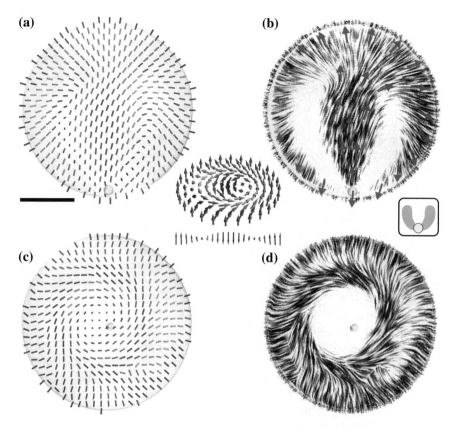

Fig. 7.3 Reconstructed director field in a droplet at $N = 2.5$ with a point defect, expelled to the surface. **a, b** The reconstructed director field and streamlines in a cross-section which includes the point defect and the symmetry axis of the droplet. The location of the point defect is marked by a yellow circle and the schematic representation of the structure is shown in the black rectangle, where the green shape represents the cholesteric bubble. **c, d** The reconstructed director field and streamlines in a cross-section which is perpendicular to the symmetry axis of the droplet. The inset shows a structure of a Bloch skyrmion (reprinted by permission from Macmillan Publishers Ltd: Nature Materials [2], copyright 2015). The scale bar is $5\,\mu\mathrm{m}$

defect is positioned in the middle of the image, and the slightly darker ring encircling the defect are the two brushes from Fig. 7.2a, seen from another perspective.

The reconstructed director in one of the planes going through the centre of the droplet and the point defect, is shown with cylinders in Fig. 7.3a and with director streamlines in Fig. 7.3b. The director streamlines are generated by projecting the director field to the plane of the cross-section as described in Sect. 5.2.1. The streamlines are not drawn where the projection of the director to the cross-section is smaller than 1/3. Therefore the director in the white areas inside the droplet in

Fig. 7.3b, d is mostly perpendicular to the plane of the cross-section. The position of the point defect in the images with director fields is marked with a yellow circle. The reconstructed director field around a point defect can be distorted because of LC fluctuations and movement of the structure on the time scale of the imaging of the 4 FCPM polarisations. Because of this, the exact location of the defect is determined from a combination of cross-sections of the FCPM intensities, I_{tot} and the reconstructed director field.

We can see that the structure is roughly axially symmetric with the point defect lying on this axis. A cross-section in a plane, perpendicular to the symmetry axis, is shown in Fig. 7.3c, d. In the centre of the droplet, the director is pointing along the symmetry axis, but it twists by $\pi/2$ to lie in the plane of this cross-section when we move towards the edge of the droplet. The in-plane director forms circular streamlines encircling the symmetry axis. We can imagine that these circular streamlines are stretching a balloon-like structure, which is pinched at the point defect to form a bell-like shape as shown in cross-section in Fig. 7.3a, b. This cholesteric formation is a motif which will frequently appear in our study of cholesteric droplets and we shall call it a cholesteric bubble (CB). By examining Fig. 7.3c, d we can see that the central part of its cross-section corresponds to the central part of a Bloch-type skyrmion shown in the central inset of Fig. 7.3, which we have also observed in the mid-plane of the homeotropic chiral nematic cell in Fig. 6.18c. The difference between the two structures appears on the edge of the structure. The vector field in a proper skyrmion twists by 2π on each line crossing its core, whereas the structure in the cross-section purely twists by π in the central part of the droplet, but on the edge the rest of the deformation is a mix of twist and bend to match the radial orientation on the surface of the droplet. The structure in the cross-section is therefore not a regular 2D skyrmion but a skyrmion embedded in a radial surrounding director. A 2D skyrmion is a topologically non-trivial texture—it has no singular points but it cannot be rearranged to uniform director with smooth deformations. This means its field has to be cut somewhere to deform it to a uniform configuration and therefore the structure is topologically protected from being erased. Because of their skyrmion-like cross-section profiles the cholesteric bubbles have increased stability too. The structure of this simple droplet with a single point defect and a cholesteric bubble can be shown schematically as in the inset to Fig. 7.3b with the yellow point marking the point defect and the two green lobes the cross-section of the cholesteric bubble.

7.2 Droplets with Layered Structures

In this section we will present the different possible layered structures we find in chiral nematic droplets with homeotropic anchoring.

7.2.1 Droplets with Bent Layers

By far the most common structure in our experiments includes a single point defect, which is expelled towards the surface of the droplet similarly as in the structure in the previous Section, but with the bulk of the droplet being occupied with bent layers. Two examples of such structures are shown in Fig. 7.4.

This is the structure, which we used for calibration of the SA algorithm in Sect. 6.3.2. Another example of the structure at lower chirality ($N = 4.7$), is shown in Fig. 7.5. The single point defect in these structures is identical to the one presented in the previous section. The point defect is pushed towards the surface of the droplet by the layered director structure which occupies most of the volume, similarly as in the droplet with the cholesteric bubble. Figure 7.5a presents the reconstructed director field of this droplet visualised with rods in a plane, which includes the only point defect in the droplet, seen at the bottom of the image. The director in the same cross-section is shown in streamlines in Fig. 7.5 to highlight the layering in the droplet. We can see that the part close to the point defect is reminiscent to a cholesteric bubble, but the lobe of the CB is spread out on one side to form an extended layer and in Fig. 7.5c we can see how the layer bends to fill the volume of the droplet. Figure 7.5d–j shows the streamlines in a series of planes which are perpendicular to those in Fig. 7.5a–c. In Fig. 7.5d we can see that the point defect has an azimuthally oriented director field in its equatorial plane. In Fig. 7.5e we can observe a similar director profile as in the case of a CB in Fig. 7.3d, but in Fig. 7.5f–h the central escape of the director (the central white area in Fig. 7.5e) is pushed to the side and the bulk of the droplet is filled with a layered twisting director field. We can observe the rotation of the director in the layers in Fig. 7.5e–j.

This type of droplets appears also at higher chiralities as we have seen in Sect. 6.3.2. With increasing N, the structure bends more to form additional layers of twist.

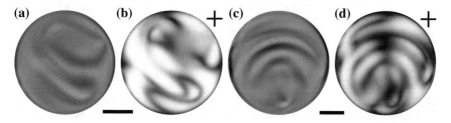

Fig. 7.4 Transmission images of droplets with a single point defect and a typical layered structure. The relative chirality of the droplets is **a, b** $N = 5.8$ **c, d** $N = 7.2$. Images **a, c** are non-polarised and **b, d** between crossed polarisers, oriented as indicated in the corners of the images. Scale bars are 5 μm

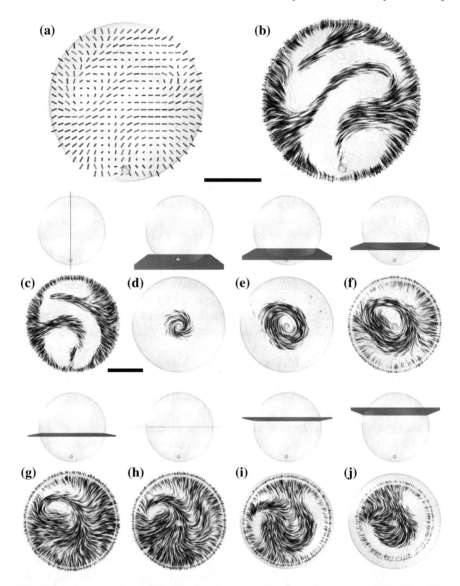

Fig. 7.5 A droplet with a single point defect and a typical layered structure at $N = 4.7$. **a** The director field in a plane with the point defect. The point defect is marked with a yellow dot and the colour of the rods corresponds to the size of the projection of the director to the cross-section. **b** The director in the same cross-section as **a**, shown in streamlines. **c** Streamlines in a plane, which is perpendicular to that from **a**, **b** and also includes the point defect. **d–j** Streamlines in a series of planes, perpendicular to the planes in (**a–c**). Scale bars are 5 μm

7.2.2 Droplets with Flat Layers

In some droplets the director can twist in a way to form flat parallel layers of cholesteric twist. Two examples of such droplets are shown in Fig. 7.6. In Fig. 7.6a–d the layers are perpendicular to the plane of the image, and the point defect which is positioned on the left edge of Fig. 7.6a is not visible. The layers of the droplet in Fig. 7.6e–h are parallel to the image plane and the point defect is visible in the bottom of Fig. 7.6g, h.

The director structure of these droplets is illustrated by the reconstruction in Fig. 7.7. Figure 7.7a shows the streamlines in a plane which is perpendicular to the plane of the flat layers and also includes the point defect close to the surface of the droplet. In this case, the reconstruction was not completely successful as can be seen in the artefacts in the image, in this case primarily by the discontinuous streamlines in the layers. Figure 7.7b shows the director streamlines in a plane parallel to the layers, which includes the point defect. We can see that the central layer is separated from the surface by another layer of twist. Figure 7.7c–h show the streamlines in a series of planes which are perpendicular to the plane of the layers. We can see how the azimuthal orientation of the director field around the point defect in Fig. 7.7c expands to form the central layer which is seen as the central white area in Fig. 7.7d. This central layer is separated by a π twist (the elliptical streamlines in Fig. 7.7d) from another layer of twist, which is the outer elliptical white area. Further along the

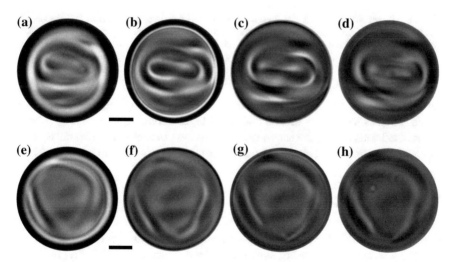

Fig. 7.6 Optical textures of droplets with a single point defect and flat parallel layers. **a–d** A series of non-polarised microscopy images of a droplet with $N = 7.2$ at different focussing depths. The point defect is in the part of the droplet which is the closest to the microscope objective, approximately in plane (**a**). **e–f** Non-polarised microscopy images of a droplet with $N = 7.5$ at different focusing depths. The point defect in the bottom part of (**g, h**). The images of each droplet are separated along the z axis by $5\,\mu$m, with the first image in the series being the closest to the objective. The scale bars are $5\,\mu$m

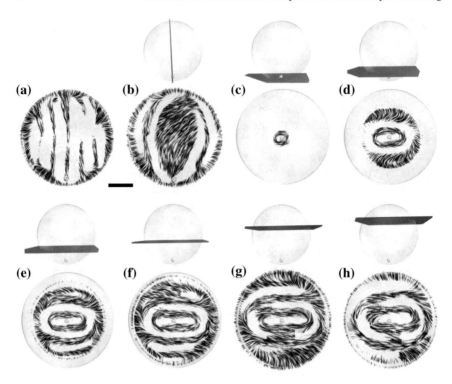

Fig. 7.7 The reconstructed director field for a droplet with a single point defect and flat parallel layers at $N = 7.2$ from Fig. 7.6a–d. **a** A cross-section, perpendicular to the plane of the layers, which includes the point defect. **b** A cross-section along a layer which starts from the point defect. **c–h** Streamlines in a series of planes which are perpendicular to the layers. The scale bar is $5\,\mu m$

axis of the droplet the director can perform even more twisting because of the bigger distance across the droplet and an additional layer of twist appears in Fig. 7.7e–h. Again, the layers are somewhat smeared in this case because of the artefacts in the reconstructed director field, but the outline of the structure can still be recognised.

7.2.3   Cylindrical Layers

One of the possible layered structures in the droplets has cylindrical symmetry. Examples of optical textures of such droplets are shown in Fig. 7.8. Figure 7.8a–d and e–h show images of two droplets at different focusing depths with a structure which appears at $N \approx 4.8-5.9$. The only difference between the two droplets is that the point defect in Fig. 7.8a–d is located at the part of the droplet, which is closest to the microscope objective and in Fig. 7.8e–h in the part which is the furthest. Figure 7.8i–l show unpolarised images of a similar droplet at higher chirality $(N = 7.5)$ with the symmetry axis at an oblique angle to the optical axis of the

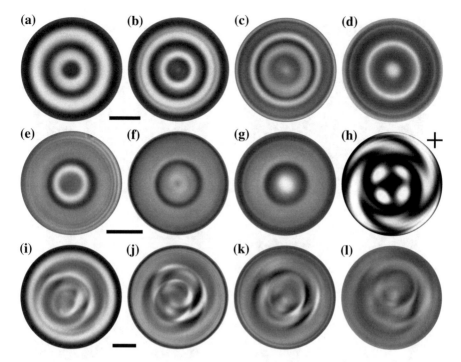

Fig. 7.8 Optical textures of droplets with a single point defect and cylindrical layers. **a–d** Non-polarised microscopy images of a droplet with $N = 4.8$. The images are taken at different locations along the symmetry axis and are separated by 4 μm. The point defect is in the part of the droplet which is the closest to the microscope objective, approximately in plane (**b**). **e–g** Non-polarised microscopy images of a droplet with $N = 4.9$, in focusing planes separated by 3 μm. The point defect is located at the far end of the droplet, approximately in the (**g**) image. **h** A polarised image of the droplet in (**e–g**) with the crossed polarisers oriented as indicated in the top right corner. **i–l** Non-polarised images of a droplet with $N = 7.5$, in focusing planes separated by 5 μm. The symmetry axis of the droplet is oriented at an oblique angle to the optical axis and the point defect is at the near end of the droplet, approximately in (**i**). The scale bars are 5 μm

microscope. This tilting breaks the symmetry of the image, but the layers can be more clearly seen in 3D with changing of the focus.

Figures 7.9 and 7.10 show the reconstructed director field of two droplets at $N = 4.8$ and $N = 7.5$. We can see that both structures really are cylindrically symmetric. The one at smaller chirality has two coaxial cylindrical layers, or in other words, the director rotates by approximately 4π across the diameter of the droplet. To be more exact, the number of rotations changes along the symmetry axis of the droplet. Close to the circular point defect with its cross-section shown in Fig. 7.9c, the director performs only a π rotation across the inner cylindrical layer (Fig. 7.9d), and then twists and bends to match the radial orientation on the surface as shown in Fig. 7.9b. Further away from the point defect, the director twists across the layers by 2π (Fig. 7.9e) and finally by 3π (Fig. 7.9f, g). In all these cross-sections the

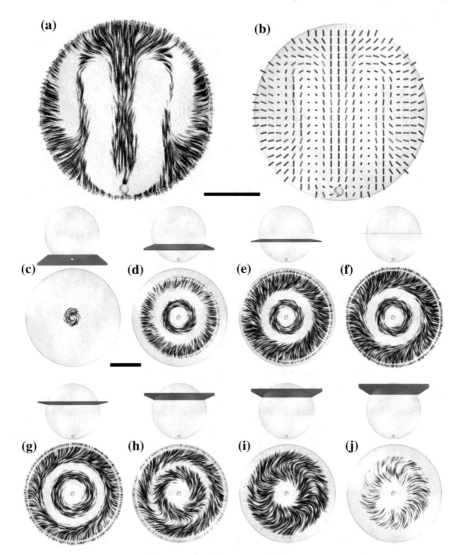

Fig. 7.9 Reconstructed director field in a droplet at $N = 4.8$ with a cylindrical layered structure. **a, b** The reconstructed director field and streamlines in a cross-section which includes the point defect and the symmetry axis of the droplet. **c–j** Streamlines in a series of cross-sections which are perpendicular to the symmetry axis. The scale bars are $5 \, \mu m$

deformation outside the outer cylindrical layer is a combination of twist and bend, so that the director matches the anchoring conditions. At the part of the droplet, furthest away from the point defect (Fig. 7.9h–j), the two layers are connected by the director bending from the inner to the outer cylindrical layer, as can be clearly seen in Fig. 7.9a, b and by the chevron-like texture in Fig. 7.9h–j.

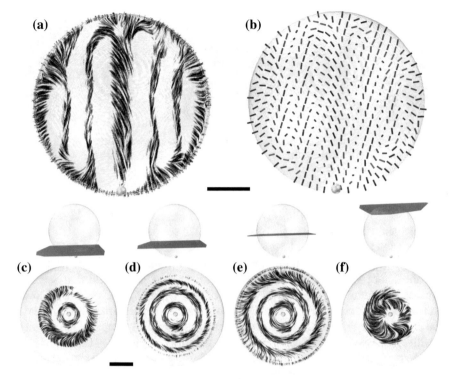

Fig. 7.10 Reconstructed director field in a droplet at $N = 7.5$ with a cylindrical layered structure. **a**, **b** The reconstructed director field and streamlines in a cross-section which includes the point defect and the symmetry axis of the droplet. **c–f** Streamlines in a series of cross-sections which are perpendicular to the symmetry axis. The scale bars are $5\,\mu\text{m}$

At higher chiralities $N \approx 6.4{-}7.8$, more layers of twist can be formed. The structure of the droplet in Fig. 7.10 is very similar to the previous one, but because of the higher chirality ($N = 7.5$), an extra cylindrical layer of twist is formed, so that the director twists by approximately 6π across the diameter of the droplet. The extra layer of twist is clearly seen in Fig. 7.10a, b, d, e, and the chevron-like texture in Fig. 7.10d indicates that the second and third layer of twist are connected with a similar bending of the director as the first and the second one in Fig. 7.10f.

7.2.4   Cholesteric Finger with a Central Hyperbolic Point Defect

In some of the quenched droplets at $N \approx 4.7{-}5.0$, the point defect is not expelled towards the surface, but instead it is stabilised in the bulk LC by layers of twist. Figure 7.11 shows a droplet at $N = 4.7$ with this structure at several different focuses.

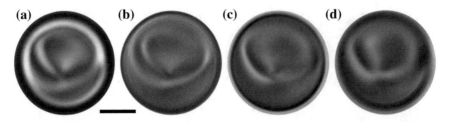

Fig. 7.11 Non-polarised transmission micrographs of a droplet ($N = 4.7$) with a central hyperbolic point defect at different focusing depths. The images are separated by $3\,\mu$m with (**a**) being the closest to the objective. The scale bar is $5\,\mu$m

The point defect is approximately in focus in Fig. 7.11b. Figure 7.12 shows the reconstructed director field in this droplet. We can see that the director field around the point defect has a hyperbolic profile in two orthogonal cross-sections in Fig. 7.12a, c and a radial profile in the third orthogonal cross-section in Fig. 7.12f. This shows that the defect is a typical hyperbolic point defect, which we presented in Sect. 2.5. We can determine the topological charge of this defect by adding arrows to the streamlines as shown in Fig. 7.12a. We would obtain a similar image if we repeated the exercise for Fig. 7.12d. We can see that the topological charge of the defect is positive, because the arrows on its symmetry axis are pointing away from the defect.

The twisted structure around the defect can be understood as a double cholesteric bubble. The top part of the droplet is almost identical to a CB, except for the missing radial point defect, where the two lobes of the CB meet. This inner CB is dressed in a layer of twist which can be understood as a second, slightly opened CB. The layered structure between the lobes of the two CBs is very similar to the structure of a looped cholesteric finger in a frustrated homeotropic cell shown in Fig. 7.12c [3]. The main difference between the two structures is that because the far-field director around the cholesteric finger is uniform, the cholesteric finger in this case doesn't include any defects, unlike the droplet where the anchoring on the surface demands a total topological charge equal to $+1$.

Figure 7.12e–k show several cross-sections perpendicular to the symmetry axis of the droplet. In the streamlines in Fig. 7.12g we can see that the lobes of the outer CB form an escaped structure which explains the reversal of the direction of arrows which point towards the defect in Fig. 7.12a. In Fig. 7.12h, i we can see, that the inner CB has the same skyrmionic cross-section as in the simpler droplets.

7.2.5 The Lyre/Yeti Structure

A structure which appears at $N \approx 5.0-6.0$ is shown in Fig. 7.13, where a point defect can be seen in the bottom of Fig. 7.13b. The structure of this droplet can be clearly observed in Fig. 7.14. We can see that a central cylindrical layered structure starts

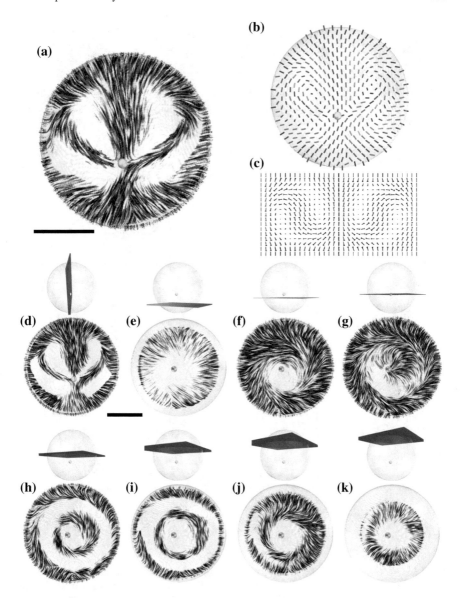

Fig. 7.12 Reconstructed director in a droplet ($N = 4.7$) with a central hyperbolic point defect. **a, b** The director and streamlines in a plane which includes the point defect and the symmetry axis of the droplet. **c** The structure of a looped cholesteric finger in a homeotropic cell, filled with a chiral nematic LC. **d** Streamlines in a plane normal to the one in (**a, b**). **e–k** A series of planes perpendicular to the symmetry axis of the droplet. The scale bars are 5 μm. Panel c reprinted from Ref. [3] by permission of Taylor & Francis Ltd

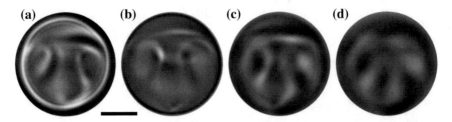

Fig. 7.13 Non-polarised transmission micrographs of a droplet ($N = 5.8$) with the Lyre/Yeti structure at different focusing depths. The images are separated by $4\,\mu m$ with (**a**) being the closest to the objective. The point defect is visible at the bottom of the image in (**b**). The scale bar is $5\,\mu m$

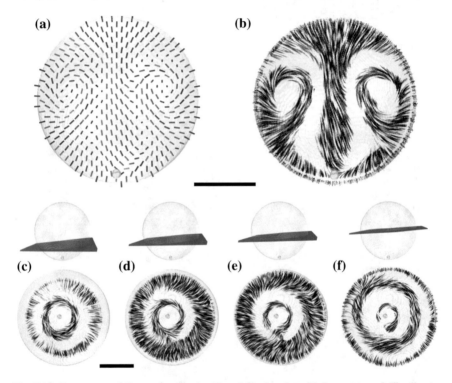

Fig. 7.14 Reconstructed director in a droplet ($N = 5.8$) with a Lyre/Yeti structure. **a, b** The director and the streamlines in a plane which includes the point defect and the symmetry axis of the droplet. **c–f** A series of planes perpendicular to the symmetry axis of the droplet. The scale bars are $5\,\mu m$

from the point defect at the bottom of Fig. 7.14a, b, extending past the centre of the droplet and then bending towards the sides of the droplet. This bent layering wraps around another cylindrical layered structure which runs around the circumference of the droplet just above the middle of the droplet as can be seen by the white patches, surrounded by circular streamlines in Fig. 7.14a, b and by the bigger ring of circular streamlines in Fig. 7.14f.

The structure is strongly reminiscent of the structures called Lyre and Yeti shown in Fig. 3.3, which were found in numerical simulations of chiral nematic droplets with planar anchoring [4], however in the presented case it is obvious that the orientation of director on the surface is homeotropic. The outline of our structure is very similar to the one of the Lyre structure, with a similar top/bottom asymmetry, with the bottom boojum from Fig. 3.3a being replaced by a 3D point defect and the top boojum not being necessary because of homeotropic anchoring. However, the circumferential cylindrical layer in Fig. 7.14b, f has a core which is well-separated from the homeotropic surface and is in that regard closer to the similar tube in the Yeti structure.

7.2.6 Droplets with Disclination Lines

Point defects are not the only type of singular areas present in cholesteric droplets with homeotropic anchoring. In some of them, closed disclination loops or ring defects can appear. In droplets with chirality $N \approx 4$ a single ring defect can appear, running around the circumference of the droplet as shown in Fig. 7.15, similarly as in small nematic droplets. In this case, the ring lies in a plane perpendicular to the image plane and is indicated by the dark straight line in both images.

Figure 7.16a–d shows FCPM intensities of a similar droplet at different polarisations in an equatorial xy plane, which is in this case perpendicular to the defect ring. Figure 7.16e shows the sum of intensities of all four polarisations I_{tot}. Areas with low I_{tot} indicate where the director is perpendicular to the xy plane. We can see that on the left and right-hand sides of the droplet, where the disclination line passes the plane at a perpendicular angle, the director quickly changes its orientation from in-plane to out-of-plane. Figure 7.16f–h show two of the fluorescence polarisations and I_{tot} in a series of planes which are perpendicular to the disclination ring as schematically indicated in the bottom row.

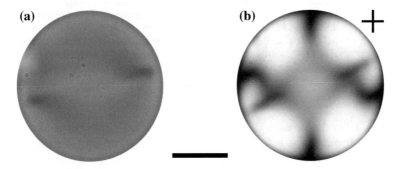

Fig. 7.15 A droplet with a ring disclination at $N = 3.9$. In this example the plane of the ring is perpendicular to the plane of the image. The left image is unpolarised and the right under crossed polarisers, oriented as indicated with the cross in the top right corner of the image. The scale bar is 5 μm

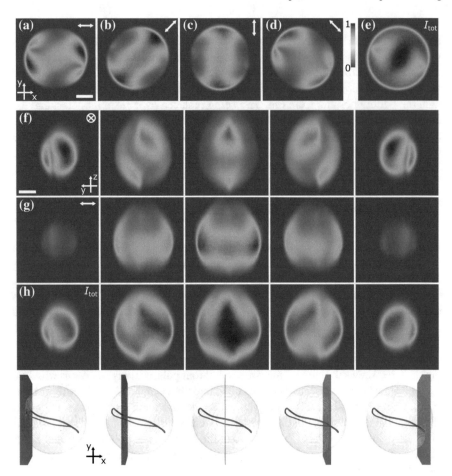

Fig. 7.16 FCPM intensities for a droplets with a ring disclination at $N = 4.4$. **a–d** FCPM intensities in an equatorial xy plane which is incidentally approximately perpendicular to the plane of the ring disclination. The polarisation in each image is marked with arrows in top right corners of the images. **e** The sum of the four FCPM intensities from (**a–d**). **f–h** FCPM intensities in a series of yz planes. The polarisation in (**f**) is perpendicular to the plane of the image, in (**g**) it is oriented as indicated in the top right corner and **h** shows the sum of all four FCPM intensities. The positions of the planes are shown schematically in the bottom row, with the approximate position of the disclination line in blue. The scale bars are $2\,\mu$m

In Fig. 7.16f the location where the disclination line passes the plane in a perpendicular direction can be seen by the sharp variation of fluorescence intensity. Such cross-sections are used to extract the location of the disclination line in different planes by hand. The points are then treated as a curve, which is numerically smoothed, and a spline is generated from the points with a POV-Ray inbuilt function called `sphere_sweep` with the option `b_spline`. The result is visualised as a ring shaped blue tube, seen in the images in the bottom row of Fig. 7.16 which

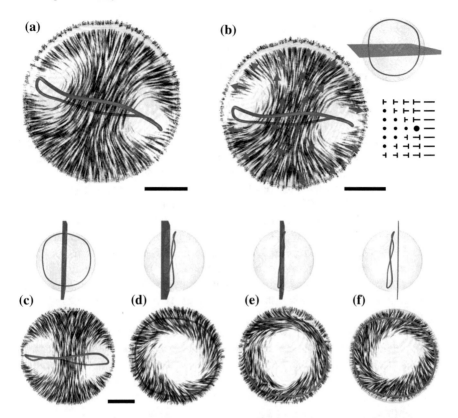

Fig. 7.17 The reconstructed director field for the droplet with a ring disclination from Fig. 7.16.
a A 3D view of the ring disclination with the director streamlines in an equatorial plane, which
is approximately perpendicular to the defect ring. **b** The director streamlines in the same plane as
shown in (**a**) with arrows to determine the sign of the topological charge of the ring disclination.
A schematic representation of the director field is shown in the inset. The bigger circle marks the
point, where the disclination line crosses this plane. The heads of the nails mark which side of the
nail is positioned above the plane. **c** The director streamlines in a plane, which is perpendicular to
the defect ring and to the planes in **a**, **b**. **d**–**f** Director streamlines in planes parallel to the plane of
the defect ring, close to the disclination. The scale bars are 2 μm

approximately corresponds to the defect ring. The resulting visualisation of the disclination line is slightly wavy. In reality the line defect is under tension and is therefore
as smooth as the surrounding director field allows it to be. Some bending of the ring is
possible because of the twisting in the chiral structure, but the short-period waviness
in the images is an artefact of the extraction of the locations of the line by hand. The
smoothing of the line reduces the waviness to give a more realistic estimation of the
shape of the ring, but the resulting line is still just an estimate of the disclination
location and is used only for visualisation purposes.

The reconstructed director field of this droplet is shown in Fig. 7.17. In Fig. 7.17a
we can see the director streamlines in an equatorial plane which is perpendicular

to the disclination ring. We can see, that the director is perpendicular to the plane of the cross-section on the inner side of the disclination line, which shows that the director on that side of the disclination is pointing along the line. This means that the disclination line has a twisted profile with the director rotating out of the plane of its cross-section to perform a π twist. A schematic representation of the director field around the disclination is shown in the inset next to Fig. 7.17b.

In Fig. 7.17b arrows are added to the streamlines to demonstrate the concept of a branch cut surface. We can see, that the arrows in the centre of the disclination loop are diverging away from the membrane stretched by the ring, which means that the defect loop is effectively equivalent to a $+1$ topological point defect.

Figure 7.17c shows the director in a plane which is perpendicular to the cross-section in Fig. 7.17a, b. We can see, that these streamlines are very similar to the ones in the previous plane with the only difference being, that the director in the centre of the droplet passes the plane of the ring approximately at an perpendicular angle. The cross-sections in Fig. 7.17d–f show the director streamlines in planes which are parallel to the plane of the disclination. The central white area in these figures indicates that the director in the centre is perpendicular to the plane of the cross-section. We can see that the director rotates by π between the edges of the droplet.

At chiralities in the range $N \approx 2.8 - 6.0$ there is enough volume in the central part of the droplet for the director to complete more twists between the diametrically opposite ends of a defect ring. Figure 7.18 shows wide-field images of examples of

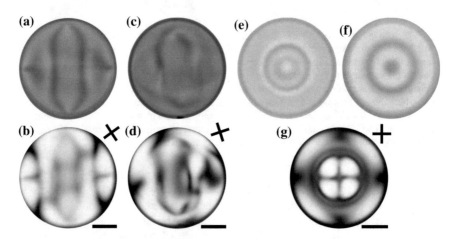

Fig. 7.18 Droplets with a ring disclination and a double twist cylinder. Top row of images is non-polarised and bottom is between crossed polarisers with the orientation of polarisers indicated in top right corner of each image. **a, b** The cylinder is oriented vertically and the ring disclination is perpendicular to the plane of the image. **c, d** The cylinder is at an oblique angle to the image plane. **e–g** The cylinder is perpendicular to the image plane. **e** and **f** are taken at different focusing planes which changes the appearance of the cholesteric layers and the point defects in the structure. The scale bars are 5 μm. Panels a and c reprinted from Ref. [1] under the terms of the Creative Commons Attribution 4.0 International License (http://creativecommons.org/licenses/by/4.0/)

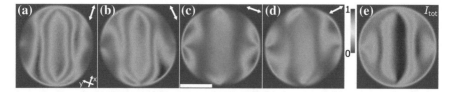

Fig. 7.19 FCPM intensities for a droplet with a ring disclination and a double twist cylinder. **a–d** FCPM intensities in a single plane for different polarisations of excitation/detection as indicated in top right corner of each image. **e** The sum of intensities at all four polarisations in this plane. The scale bar is 5 μm

such droplets and Fig. 7.19 shows FCPM intensities in the plane which includes the two point defects. In these droplets the central part is filled with a doubly twisted cylindrical structure which terminates in a point defect on both ends where the cylinder meets the surface of the droplet. Both of these defects are of the same type as the defects close to the surface in the previous droplets. Because a line defect is present in the droplet, the point defects cannot be unambiguously assigned a topological charge, because the sign of the charge depends on the choice of the branch cut surface of the defect ring. This choice also changes the topological charge of the defect loop. In such situations we can only determine the total topological charge modulo 2 or in other words, if the total charge is odd or even [5–7]. Because of the homeotropic anchoring the total topological charge of the droplet must be odd (or exactly $+1$ for any choice of a branch cut surface). The topological charge of the line defect must also be odd, because it has a half-integer winding number of its cross-section and the proximity of the surface forces it to have a constant cross-section [5]. The total topological charge of the droplet therefore really is odd, because the topological charge of the two point defects with $|q| = 1$ is even.

The detailed behaviour of the director in the central part of the droplet is shown in Fig. 7.20b. Here we can see, that the director around the disclination has a twisted profile, similar to the one shown in Fig. 2.9d. If we traverse the droplet between the two opposite parts of the disclination line, we can see that the director has a constant direction of rotation and rotates by 3π. This rotation is presented in more detail in Fig. 7.20c–e, which show the streamlines in cross-sections which are perpendicular to the symmetry axis of the droplet. The central part of each image has a skyrmionic director profile, similar to the cross-sections of cholesteric bubbles. One of the differences between this structure and the cholesteric bubbles is that here, the rotation does not stop before reaching 2π as it does in CBs. Also, the twisted layers which terminate without a singularity in a CB, are here extended so they terminate at another point defect. A *cholesteric cylinder* (CC), as we will call such a structure, is therefore in its simplest form bound to two point defects of the same type.

At even higher chiralities ($N = 5.6$–6.4), more than one cholesteric cylinder can appear inside the ring. Figure 7.21 shows non-polarised transmission images of two slightly different droplets with two cholesteric cylinders encircled by a ring disclination. In the smaller droplet ($N = 5.6$; top row) both cholesteric cylinders lie

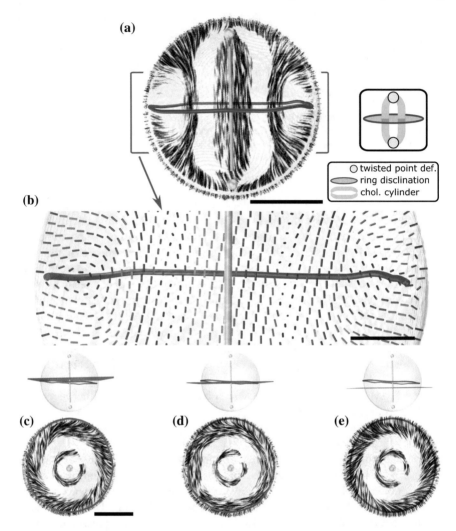

Fig. 7.20 Reconstructed director field of a droplet with a ring disclination and a cholesteric cylinder between two point defects. **a** The director streamlines in a plane which includes the two point defects and dissects the cylindrical layer. **b** The enlarged central part of the droplet around the disclination line shows the full director field. Here we can clearly follow the rotation of the director field in a direction perpendicular to the axis of the cylinder. **c–e** A series of cross-sections perpendicular to the cylinder, close to the ring disclination. The director on the edge of the droplet in (**c**) is spiralling in one direction, in (**d**) which is very near to the ring disclination it has a changing profile and in (**e**) the direction of spiralling is reversed compared to (**c**). The rod is added to help visualise the spatial relations of the point defects. The scale bars in (**a, c–e**) are 5 μm and in (**b**) 2 μm

at an oblique angle to the image plane and in the bigger one ($N = 6.4$; bottom row), the left cylinder is approximately in the plane of the image, and the other one is at an oblique angle. The ring disclination is in both droplets approximately perpendicular

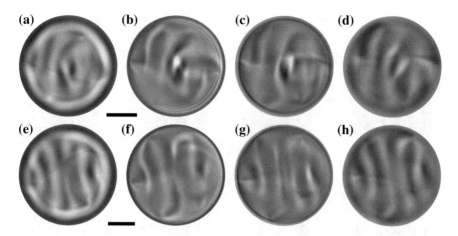

Fig. 7.21 Droplets with a ring disclination and two cholesteric cylinders. In the top row, the images show different focuses separated by 4.5 μm of a droplet with $N = 5.6$ and in the bottom row of a droplet with $N = 6.4$. In both droplets the approximate plane of the disclination ring is perpendicular to the plane of the image. In the top row the two cylinders are at an oblique angle to the image plane and in the bottom row, one of them is approximately in the image plane and the other one is oblique. The scale bars are 5 μm

to the plane of the image and therefore different sections of the disclination are in focus in each image. We can see, that the disclination does not lie in a plane but runs around the two cylinders in a winding way.

The FCPM images of the droplet with $N = 6.4$ in Fig. 7.22 reveal more details about the structure. The different polarisations in Fig. 7.22a–e show the orientation of director in an xy plane which includes one of the cylinders. The polarisation in Fig. 7.22a reveals the two defects in the left cylinder and how the director in its core is aligned with the axis of the cylinder. In the right part of the image we can see the outline of a cross-section of the other cylinder and that the two cylinders are separated by layers of twist. The intensity in Fig. 7.22c shows the cross-sections of the disclination line on the left and right edges of the droplet and Fig. 7.22e shows, where the director lies in the plane of the image.

Figure 7.22f–j show the intensities for the different polarisations in a vertical yz plane, which is perpendicular to the left cylinder. The intensity in Fig. 7.22f, which is orthogonal to the image plane, shows the cylindrical symmetry of both cylinders and that the director in their cores is pointing along their axes. We can also see, how the two cylinders are separate from each other. In Fig. 7.22h, j we can see the locations of the disclination line in several locations, which are also seen in Fig. 7.22j.

Figure 7.22k–m present intensities of two orthogonal polarisations and I_{tot} in a series of vertical xz planes which are approximately perpendicular to the approximate plane of the disclination line. In Fig. 7.22l, m we can nicely follow the position of the disclination. In the first and the last image in the series, the disclination line is indicated by the sharp drop of intensity, because the director at the disclination

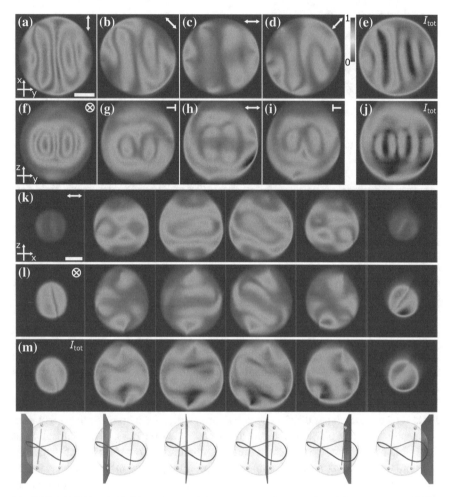

Fig. 7.22 FCPM intensities for a droplet with a ring disclination and two cholesteric cylinders ($N = 6.4$). **a–d** FCPM intensities in a single xy plane for different polarisations of excitation/detection as indicated in top right corner of each image and **e** the sum of intensities at all four polarisations in this plane. **f–i** FCPM intensities in a single yz plane for different polarisations of excitation/detection as indicated in top right corner of each image and **j** I_{tot} in this plane. **k, l** Intensities for two orthogonal polarisations and **m** I_{tot} in a series of vertical planes, which are orthogonal to the disclination line. The positions of the planes are indicated in the bottom row. The scale bars are 5 μm

rotates from the radial orientation, which is apparent as the bright surrounding of the disclination, to a vertical orientation, which is perpendicular to the probing polarisation and therefore has low intensity. The effect is inverse in the middle four images of the series. The surrounding of the disclination on the top and bottom of the droplet is again radial, so the vertical director has low intensity both in the single polarisation Fig. 7.22l and in the total FCPM intensity in Fig. 7.22m. At the disclination, the

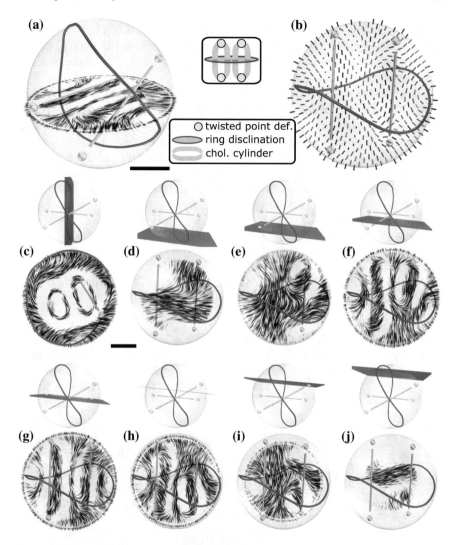

Fig. 7.23 Reconstructed director field of a droplet with a ring disclination and two cholesteric cylinders ($N = 6.4$). **a** The director streamlines in a plane which includes two point defects and the axis of one of the cylinders. **b** The director field in the plane from (**a**). **c** The streamlines in a plane, which is perpendicular to the one in (**a, b**). **d–j** A series of cross-sections perpendicular to the one in (**c**). The rods are added to help visualise the spatial relations of the point defects. The scale bars are 5 μm

director rotates to the y direction to point along the disclination so one side of the disclination line appears bright.

The reconstructed director in Fig. 7.23 confirms that each of the two cylinders has the same structure as in the simpler droplet. A 3D representation of the structure in Fig. 7.23a shows how the two cylinders are positioned at an oblique angle relative to the ring and Fig. 7.23b shows the director field in a plane which includes the axis of one of the cholesteric cylinders and dissects the other one. We can see, that the between two opposite ends of a disclination line across both cholesteric cylinders the director twists by 5π.

The plane in Fig. 7.23c is orthogonal to the ones in Fig. 7.23a, b and shows the cylindrical symmetry of the two cholesteric cylinders. The series of cross-sections in Fig. 7.23d–j are approximately perpendicular to the plane of the disclination ring, and we can follow the location of the disclination from a different perspective than in Fig. 7.22k–m. We can see that the director on the inner side of the disclination is always pointing along the disclination, which means that the disclination has a twist profile of the director cross-section everywhere in the droplet. The images Fig. 7.23d–j also show that the disclination is everywhere separated from the centres of the cholesteric cylinders at least by a π twist, which is indicated by the small white areas at locations of the disclination in Fig. 7.23e–i.

7.2.7 Combinations of Cholesteric Bubbles and Cholesteric Cylinders

In some situations cholesteric bubbles and cholesteric cylinders can appear in the same droplet. An example of this is a structure with 3 point defects which can appear at $N \approx 4.7-6.0$. Examples of wide field images of droplets with such a structure are shown in Fig. 7.24. We can see that the optical texture of this structure is very similar

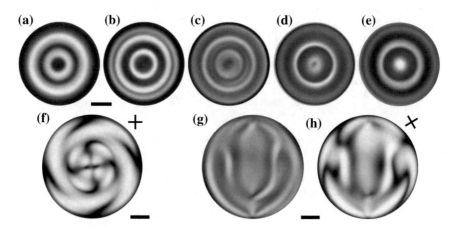

Fig. 7.24 A droplet with a combination of a cholesteric bubble and a cholesteric cylinder. Images (**a–e**) and (**g**) are non-polarised and (**f, h**) are between crossed polarisers with the orientation of polarisers indicated in top right corner of each image. **a–e** A series of slices at different focuses separated by $4\,\mu$m in a droplet ($N = 5.7$) with the symmetry axis oriented vertically. **f** An image of a droplet ($N = 5.6$) with the same orientation as in (**a–e**) under crossed polarisers. **g, h** A non-polarised (**g**) and an image under crossed polarisers (**h**) of a droplet ($N = 5.7$) with the symmetry axis in the plane of the image. The scale bars are $5\,\mu$m

to the one of the structure with cylindrical layers in Fig. 7.8 and of the cholesteric cylinder with the ring disclination in Fig. 7.18 when viewed along their symmetry axis. The easiest way to distinguish these textures is under crossed polarisers. The structure with the ring disclination has a dark cross with strait brushes and a well-defined dark ring as seen in Fig. 7.18d, in the structure with cylindrical layers the dark brushes have a V-shaped kink (Fig. 7.8h) and in the present case, the dark ring is blurred and the brushes are spiralling outwards without a kink (Fig. 7.24f).

We can get more information about the structure of this droplet from images where the symmetry axis is lying in the plane of the image (Fig. 7.24g, h). We can notice the locations of three point defects and outlines of structures which are similar to the ones in the previous droplets. The point defect in the bottom of the image is attached to a structure, reminiscent of a cholesteric bubble, and the shape between the top two point defects looks a lot like the cholesteric cylinder in Fig. 7.18a.

The reconstructed director field shown in Fig. 7.25 confirms this. The top two point defects in Fig. 7.25a terminate the doubly twisted cylinder which is aligned with the symmetry axis of the CB, which is located in the bottom of the image. The CB is, as in all the previous cases, anchored on a $+1$ point defect close to the surface of the droplet, but in this structure the hyperbolic defect which usually sits just outside the CB is pushed deep between the lobes of the CB by the cholesteric cylinder. The lobes of the CB are compressed between the surface of the droplet and the sides of the cylinder to form layer-like areas of twisting director in which the cholesteric cylinder sits. The structure is shown schematically in the inset to Fig. 7.25a and Fig. 7.25b shows the full director field in which we can clearly see the rotation of the director field between the different layers.

Figure 7.25c–j show the director streamlines in a series of planes which are perpendicular to the axis of the droplet. The bottom point defect in Fig. 7.25c has a twisted director profile, which evolves into an azimuthal orientation inside the lobes of the cholesteric bubble shown in Fig. 7.25e. The -1 point defect in Fig. 7.25f has a radial director profile in its equatorial plane, and above it in Fig. 7.25g–i the director forms the second, inner layer of twist, which is the cholesteric cylinder. The outer ring of azimuthally oriented director in Fig. 7.25g is the cholesteric bubble. The second $+1$ defect shown in Fig. 7.25j has a similar twisted director field as the first one in Fig. 7.25c.

Because one end of the CC is no longer connected to the surface of the droplet, this influences the structure of this bulk defect. In Fig. 7.25f we can see it acquires a hyperbolic arrangement of the surrounding director field in the cross-section which includes its symmetry plane and in Fig. 7.25f which is in the mid-plane of the defect, the director is radial. The sign of topological charge of this defect is opposite compared to the previous point defects in cholesteric cylinders. Because the outgoing vectorised director from the surface $+1$ defect in the cylinder connects to the hyperbolic defect without crossing a direction changing membrane, the hyperbolic defect is assigned a -1 charge as seen in Fig. 7.25a. The cholesteric cylinder in this structure is therefore terminated by two mismatched point defects which is different from the CCs in droplets with ring defects. The structure in the droplet is schematically shown in the inset to Fig. 7.25.

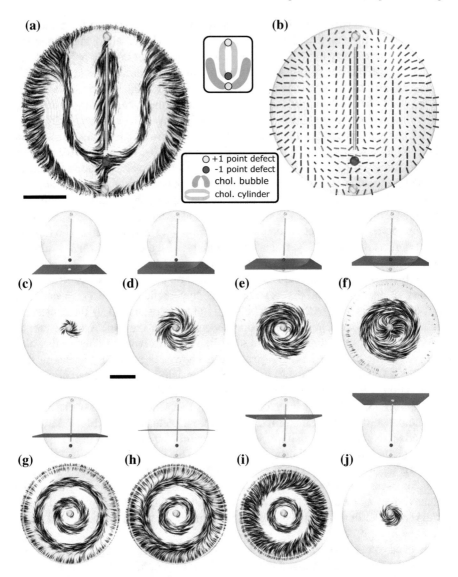

Fig. 7.25 Reconstructed director field of a droplet with a cholesteric bubble and a cholesteric cylinder from Fig. 7.24a–e ($N = 5.7$). **a** The director streamlines and **b** the director field shown in cylinders in a plane which includes the symmetry axis of the droplet and all three point defects. The streamlines show the outline of the structure as highlighted in the inset. **c–j** A series of cross-sections perpendicular to the symmetry axis of the droplet. The rods are added to help visualise the spatial relations of the point defects. The scale bars are 5 μm

A similar structure shown in Fig. 7.26 appears at $N \approx 6.4–6.9$. In Fig. 7.26a, b the structure is oriented with its symmetry axis in the image plane and onion like layers can be seen with five point defects along the symmetry axis of the structure.

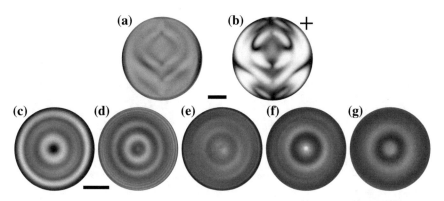

Fig. 7.26 A droplet with an onion-like structure. **a, b** A droplet at $N = 6.9$ with the symmetry axis in the plane of the image. A non-polarised image is shown in (**a**) and in (**b**) under crossed polarisers, with their orientation indicated in the top right corner. **c–g** A series of slices at different focuses separated by 4 µm in a droplet ($N = 6.4$) with the symmetry axis oriented vertically. The point defect in the cholesteric bubble is in the plane closest to the objective, approximately in (**c**). The scale bars are 5 µm. Panel a reprinted from Ref. [1] under the terms of the Creative Commons Attribution 4.0 International License (http://creativecommons.org/licenses/by/4.0/)

The locations of point defects can be most clearly seen at each point where a pair of petal-shaped bright patches meet in Fig. 7.26b. Figure 7.26c–g shows a series images in different focusing planes for a droplet with its symmetry axis along the z direction. The several layers of twist are clearly seen as the concentric rings. The point defects are not visible as they obscure each other along the symmetry axis, but we can see that the brightness of the central part of the droplet in the different focusing planes varies strongly.

The reconstructed director field in Fig. 7.27 reveals the onion-like layered structure in the top part of the droplet which is nested in a cholesteric bubble, located in the bottom of the image. Unfortunately the reconstructed director field in Fig. 7.27a shows many artefacts, which can be seen in Fig. 7.27b as areas where the layers are interrupted. This is a consequence of the relatively short pitch and lensing by the cholesteric layers of the structure, which affected the FCPM intensities. Even with these artefacts the general features of the structure can be still recognised.

The onion-like part of the structure has two concentric spherical layers which can be nicely seen in Fig. 7.27e. The inner layer can be understood as the cholesteric cylinder from the the previous structure, which has been compressed a bit along its axis. The additional layer of twist includes an additional pair of point defects and together with the point defect in the CB, the five defects are arranged collinearly along the symmetry axis of the droplet.

All 5 point defects have their symmetry axes aligned along the symmetry axis of the droplet, which means that the signs of their topological charge alternate. The equatorial cross-sections of the two point defects close to the surface can be seen in Fig. 7.27c, j and are similar in shape to surface point defects in other layered droplets. Both of them carry a +1 topological charge because of the proximity of the

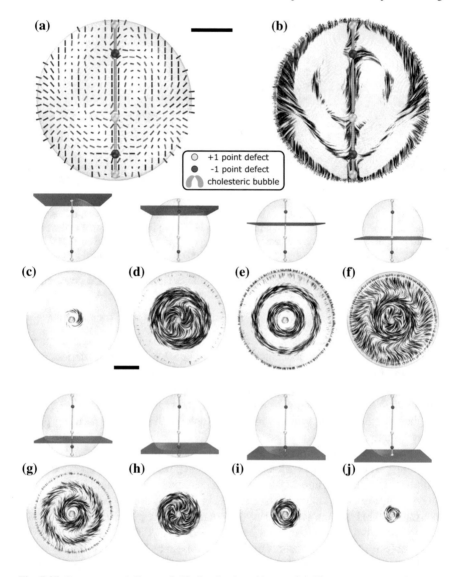

Fig. 7.27 Reconstructed director field of a droplet with an onion-like structure ($N = 6.5$). The director field in a plane which includes the symmetry axis is shown in cylinders in (**a**) and in streamlines in (**b**). The locations of point defects are marked with yellow dots for defects with $+1$ and with magenta dots for defects with -1 topological charge. **c–j** A series of cross-sections perpendicular to the symmetry axis of the droplet, with (**c**), (**d**), (**f**), (**h**) and (**j**) showing cross-sections through point defects. The rods are added to help visualise the spatial relations of the point defects. The scale bars are $5\,\mu$m

surface of the droplet. In Fig. 7.27b we can see that the director cross-sections of the middle 3 point defects in a plane which includes their symmetry axis is hyperbolic for all three defects. Figure 7.27e, f, h shows their equatorial cross-sections which are radial in all three cases. This means all middle three defects have a hyperbolic configuration of director field but nonetheless carry different topological charges.

7.3 Strings of Point Defects

Some of the droplets do not exhibit the typical cholesteric layering but instead form multiple cholesteric bubbles, which serve as barriers between point defects and prevent their annihilation. The point defects arrange in string-like configurations and alternate between $+1$ and -1 topological charges, where every $+1$ defect is close to the droplet surface and the -1 defects are in the bulk of the LC.

The simplest of these structures appears at $N \approx 2.7-4.9$ and is formed of 3 point defects and 2 cholesteric bubbles. Figure 7.28 shows several wide field images of different orientations of such a droplet. As we can see this structure can appear in various optical textures because different orientations of the structure result in different projections of the point defects and cholesteric bubbles. The line-like structures seen in the non-polarised images correspond to areas where the director field is twisted

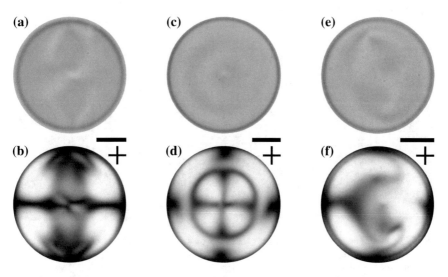

Fig. 7.28 Optical textures of droplets with three collinear point defects. **a, c, e** show non-polarised and **b, d, f** polarised microscopy images between crossed polarisers, oriented as indicated in the top right corner of each image. The line on which the defects lie is (**a, b**) in the plane of the image, **c, d** perpendicular to and centred in the middle of the image and in (**e, f**) at an oblique angle to the image. The scale bars are 5 μm. Panels **a, c** and **e** reprinted from Ref. [1] under the terms of the Creative Commons Attribution 4.0 International License (http://creativecommons.org/licenses/by/4.0/)

in a different direction than its surroundings. These are the lobes of the CBs. Point defects can be seen as areas where these structures are pinched or where the intensity in the polarised images strongly varies around. In Fig. 7.28c, d only one point defect is visible because the other 2 are obscured by it as they lie on the same axis which is perpendicular to the plane of the image. The two images are centred on the equatorial plane of the droplet, which includes the middle point defect. The outline of an out-of-plane cholesteric bubble can be seen as a dark ring surrounding the central point defect.

In Fig. 7.28e, f the defects lie on a line at an oblique angle to the plane of the image and the outlines of the cholesteric bubble are smeared. The matching of the structures in droplets with such textures was also confirmed by simulations presented in Ref. [8]. In the study the director of a droplet with 3 collinear defects reconstructed by our method served as the structure, the orientation of which was varied to study its effect on the appearance of optical textures under crossed polarisers. The study found close agreement between experimental and simulated images.

Figure 7.29 shows several FCPM cross-sections for two differently orientated droplets from which the positions of 3 point defects and two cholesteric bubbles can be clearly resolved. The centres of the lobes of the CBs are most clearly seen in Fig. 7.29e as the blue areas which indicate director with normal orientation to the xy plane and in Fig. 7.29h as the red areas which because of the probing polarisation being perpendicular to the plane, indicate the out-of-plane orientation of director. In both images the separation of the two bubbles by an area of director twisted to the in-plane orientation is obvious.

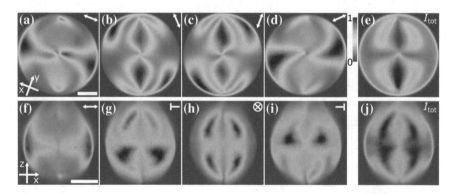

Fig. 7.29 FCPM intensities for two cross-sections of a droplet with three collinear defects. **a–e** show FCPM intensities in a equatorial plane in which lie all three defects. The polarisation of excitation/detection is marked in the top right corner of each panel. Panels **f–j** show FCPM intensities in a vertically oriented plane with all 3 defects for a droplet with vertically oriented symmetry axis. **e, j** show the sum of all four polarisations I_{tot} in the respective plane which is proportional to the projection of the director to the xy plane. The scale bars are 5 μm. Panels **a–d** reprinted from Ref. [9] and panels **f–i** reprinted from Ref. [1], both under the terms of the Creative Commons Attribution 4.0 International License (http://creativecommons.org/licenses/by/4.0/)

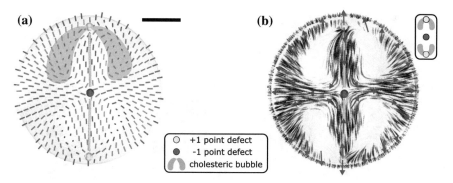

Fig. 7.30 Reconstructed director field of the droplet with 3 collinear defects. **a** shows the director in an equatorial plane, which includes all three point defects. The green area indicates the position of one of the two cholesteric bubbles. **b** The director field from (**a**), shown in streamlines. The streamlines are not drawn where the projection of the director field to the cross-section is smaller than 1/3. The white areas therefore indicate areas of out-of-plane director field which in this case correspond to the cholesteric bubbles as shown in the schematic representation in the inset. The red arrows mark the direction we assigned to the director field according to the convention that the field on the surface is pointing outwards. The scale bar is 5 μm. The rods are added to help visualise the spatial relations of the point defects. Reprinted from Ref. [9] under the terms of the Creative Commons Attribution 4.0 International License (http://creativecommons.org/licenses/by/4.0/)

The positions of CBs are also confirmed in the reconstructed director field in Fig. 7.30a and identified as the areas without the streamlines in Fig. 7.30b. The two CBs in this droplet retain their cylindrical symmetry and are still anchored on a +1 point defect close to the surface of the droplet. Between the two CBs there appears a new point defect with a different geometry than the two +1 defects. By examining the director field around this defect in Fig. 7.30a, b we can clearly see it is a hyperbolic point defect. It is easy to see that the shape of the director field surrounding the hyperbolic point defect nicely matches the director field around the CB so it can be embedded neatly between the two CBs with symmetry axes of all the objects aligned.

We can find the topological charge of the defects by prescribing orientation to the headless director field and observing the behaviour of the vectorised director on a sphere surrounding a point defect. In the case of a droplet with 3 collinear point defects, it is enough to assign direction just in one plane, as the structure is cylindrically symmetric. If we choose the director on the surface of the droplet to point outwards, we find that the two twisted point defects close to the surface really do have a topological charge of +1. This is apparent in Fig. 7.30b from the director on their symmetry axis pointing away from the point defect. The central hyperbolic defect also lies on the same axis and the director on its symmetry axis points inwards, which gives it a −1 topological charge. The total charge of the droplet is therefore $q = 2 \times (+1) + (-1) = +1$ as predicted by Poincaré-Hopf theorem for a droplet with homeotropic anchoring.

The cross-sections in Fig. 7.31 illustrate the skyrmionic nature of the cholesteric bubbles. Figure 7.31a shows the strongly twisted director field in the equatorial plane

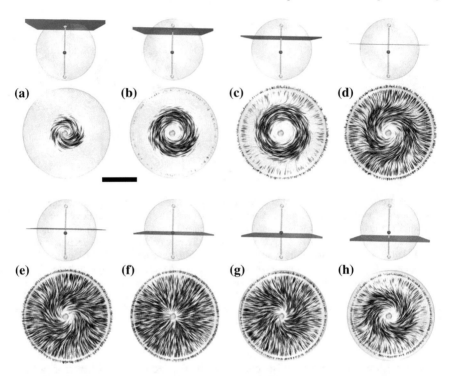

Fig. 7.31 Reconstructed director field of the droplet with 3 collinear defects in cross-sections, perpendicular to the symmetry axis of the droplet. Image **a** shows the director in an equatorial plane of the top point defect. **b–d** The director field in different cross-sections of the cholesteric bubble. **e–h** The director in cross-sections close to the −1 defect. The scale bar is 5 μm

of one of the two +1 defects on the surface of the droplet. Figure 7.31b, c shows the planes in which the CB has a director profile which strongly resembles a Bloch 2D skyrmion by twisting by almost π from the direction in the core of the bubble, which is aligned with the axis of the structure to the direction in the region close to the surface of the droplet. Because this twist happens in any direction away from the axis, this part of the CB is a double-twist cylinder. Further towards the central point defect the director still twists away from the axis of the structure but tilts less out-of-plane closer to the surface of the droplet as seen in Fig. 7.31d, e. In the equatorial plane going through the central −1 defect the director field is completely radial and virtually dissects the droplet into two halves, each containing one CB. The mid-plane of a hyperbolic point defect is radial because of the cylindrical symmetry of the point defect, but Fig. 7.31e, g demonstrates that the director profile around the defect is deformed because of twisting of the director. In planes which dissect the other CB in Fig. 7.31g, h, the director profile starts to twist in the opposite direction because of the inverted orientation of the other CB, but still preserves the handedness of the structure.

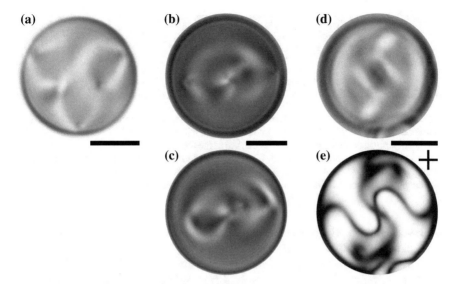

Fig. 7.32 Optical textures of droplets with 5 point defects in a V-shaped constellation. **a** shows a non-polarised microscopy image of a droplet with all the defects in an equatorial plane. **b, c** a droplet with the V-shaped constellation in a vertical plane. **b** is focused on the lower arm with 3 point defects and an outline of an out of plane cholesteric bubble is visible as a bright circle. **c** is focused above (**b**), approximately on the top −1 defect, seen as a dark point a bit right from the centre of the droplet. **d, e** A droplet with a vertically oriented V-shaped constellation, the +1 point defect at the bottom of the V is oriented towards the objective. The image under crossed polarisers in (**e**) reveals positions of the three +1 defects—the defect at the bottom of the V appears in the middle of the picture as a dark twisted cross, and the two +1 defects at the ends of the arms of the V can be seen on the top and bottom of the picture, similarly as in Fig. 7.28b. The scale bars are 5 μm. Panel a reprinted from Ref. [9] under the terms of the Creative Commons Attribution 4.0 International License (http://creativecommons.org/licenses/by/4.0/)

With increasing relative chirality even more pairs of ±1 defects are added, each separated by a cholesteric bubble anchored on a +1 point defect close to the surface of the droplet. At $N \approx 2.9$–5.9 a third cholesteric bubble with an associated pair of defects appears and together with the other bubbles forms a V-shaped structure. Figure 7.32 shows wide field images of different orientations of such a structure. We can see that again the appearance of the droplet strongly depends on the orientation of the structure. In Fig. 7.32a all 5 defects are visible in a single focusing plane with the outlines of the cholesteric bubbles. In a differently oriented droplet in Fig. 7.32b, c one of the arms of the V lies in a single plane which is focused in Fig. 7.32b, but the focus needs to be changed to observe other point defects (Fig. 7.32c). With transmission microscopy also out-of-focus parts of the structure contribute to the image, for example the bright ring in Fig. 7.32a which is caused by the cholesteric bubble at the top of the droplet. With different orientations of the droplets one or more defects can be obscured by other parts of the structure, for example only 3 defects can be clearly seen in Fig. 7.32d, e as the −1 defects are hidden by the top

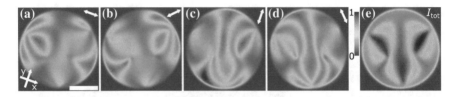

Fig. 7.33 FCPM intensities for a cross-section of a droplet with 5 point defects in a V-shaped constellation. The panels show FCPM intensities in an equatorial plane in which lie all 5 defects. The polarisation of excitation/detection is marked in the top right corner of each panel. **e** shows the sum of all four polarisations I_{tot} which is proportional to the projection of the director to the xy plane. The scale bar is 5 μm. Panels **a–d** reprinted from Ref. [9] under the terms of the Creative Commons Attribution 4.0 International License (http://creativecommons.org/licenses/by/4.0/)

cholesteric bubble which is connected to the point defect which forms the twisted cross in the centre of the image.

Figure 7.33 shows FCPM intensities in a xy plane in which lie all 5 point defects. Positions of point defects can be resolved from the individual excitation/detection polarisations in Fig. 7.33a–d as areas where the intensity distribution appears pinched. The sum of all four intensities I_{tot} in Fig. 7.33f indicates the out-of-plane twisting of the director which corresponds to the cores of the lobes of the cholesteric bubbles.

In the reconstructed director field in Fig. 7.34a we can recognise the locations of the point defects in this plane by the director field pattern around them and the streamline representation in Fig. 7.34b reveals the outlines of CB as the areas of out-of-plane director. Schematically the structure can be represented with CBs as shown in the inset to Fig. 7.34b. Just as in the simpler two droplets with CBs, each CB is anchored on a $+1$ point defect close to the surface of the droplet. Additional volume of the droplet allows the formation of another cholesteric bubble, with two associated point defects—a $+1$ defect close to the surface on which the additional CB is anchored on and another hyperbolic defect which separates the additional CB from one of the previous ones. It is obvious that if the defect close to the surface has $+1$ charge, the hyperbolic defect must have -1 topological charge to ensure the total charge in the droplet. This can be verified by tracing the orientation of the vectorised director along the symmetry axes in Fig. 7.34b—each of the point defects, on which the CBs are anchored on has diverging director arrows along its symmetry axis and for the hyperbolic ones they are converging towards the defect, similarly as in the previous droplet. The direction of the arrows inverts when we pass any of the lobes of the cholesteric bubbles by first twisting the director out of the plane of the cross-section and continuing the twist to complete a π rotation. There is a difference in positioning of the hyperbolic defects compared with the droplet with 3 point defects and 2 CB. In the previous droplet the hyperbolic defect in the centre was positioned on the symmetry axis of both CBs and all symmetry axes were aligned. In the droplet with 3 CBs, the hyperbolic point defects lie off the symmetry axis of the ending CBs, and they are not aligned with the symmetry axis of the central CB—both of the hyperbolic defects are moved off-centre and simply leaned against the lobe

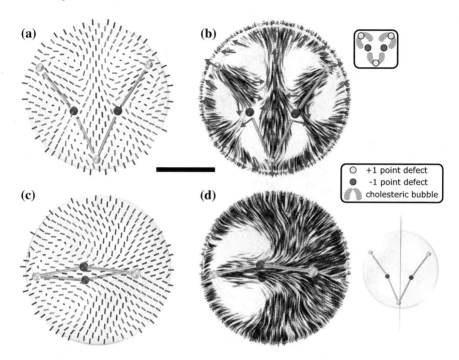

Fig. 7.34 Reconstructed director field of the droplet with a V-shaped constellation of 5 point defects. **a** shows the director in an equatorial plane, which includes all 5 point defects. **b** The director field from (**a**), shown in streamlines. The white areas indicate areas of out-of-plane director field which in this case correspond to the cholesteric bubbles as shown in the schematic representation in the inset. **c** The director profile in a vertical cross-section in the middle of the central CB. **d** The director field from (**c**), shown in streamlines. These are 3D visualisations of the structure, so the size of the point defects is inversely correlated with the distance from the camera. The scale bar is 5 μm. Panels **a** and **b** reprinted from Ref. [9] under the terms of the Creative Commons Attribution 4.0 International License (http://creativecommons.org/licenses/by/4.0/)

of the central CB, so that their hyperbolic surrounding director field continuously couples to that of the central CB. We can see from this that the CBs are not coupled to hyperbolic defects in a fixed way, but can move relatively to them, giving some flexibility to the strings of point defects.

The vertical cross-section in Fig. 7.34c, d dissects the central cholesteric bubble and shows it retains the cylindrical symmetry it had in the simpler droplets with 1 or 3 point defects. We can see that the director field further away from the CB becomes mostly in-plane and almost radial, seemingly separating the droplet in two parts.

Figure 7.35 shows streamlines in a sequence of cross-sections dissecting the droplet as indicated in the insets. Figures 7.35a, e and 7.34c, d demonstrate that the central CB in this structure retains the cylindrical symmetry which it had in droplets with 1 or 3 point defects. Further slices in Fig. 7.35b show how the two hyperbolic defects terminate the CB, the core of which diverges at that point to form

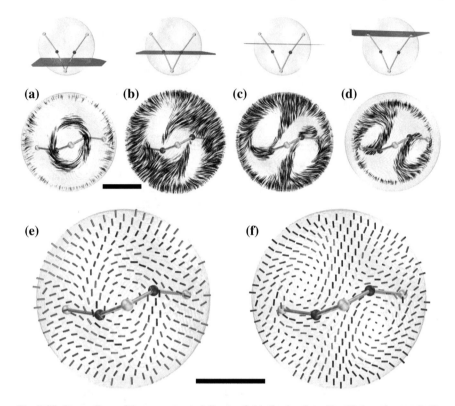

Fig. 7.35 Streamlines of the reconstructed director field of a droplet with a V-shaped constellation of 5 point defects. **a–d** Streamlines of the director field in several planes as indicated in the insets. **e, f** The director field shown with cylinders in the same planes as (**a**) and (**c**), respectively. These are 3D visualisations of the structure, so the size of the point defects is inversely correlated with the distance from the camera. The scale bars are $5\,\mu$m

the central layer seen in Fig. 7.35c, f and d. In this part of the structure between the side arms of the V-shaped constellation the director has more volume to complete several twists across the diameter—it rotates by 3π in this case.

Figure 7.36 shows a series of cross-sections along one of the arms of the droplet. In Fig. 7.36a we can see the CB on the terminal $+1$ point defect also has cylindrical symmetry. Figure 7.36b presents the director field in a plane going through the hyperbolic point defect. We can see the director is mostly in-plane similar to the one in the droplet with 3 point defects, again apparently separating the CB from the rest of the droplet. Figure 7.36c shows the quasi-layered structure in the bulk of the droplet, the twisting of which can be studied in detail in the vector representation in Fig. 7.36e. The last slice Fig. 7.36d, f shows the director streamlines in a plane which includes the three defects in the other arm of the constellation. The rough outlines of the CBs can be recognised, but their structure is deformed because of the proximity of the homeotropic boundary and the merging of the twisted regions. Because of this

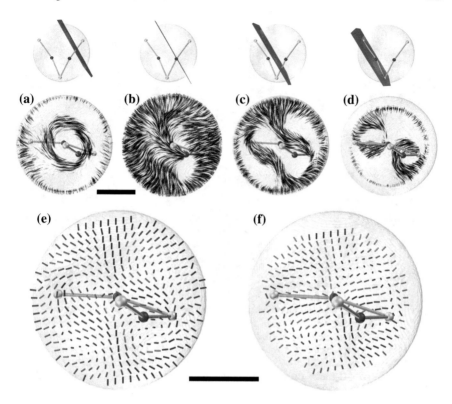

Fig. 7.36 Streamlines of the reconstructed director field of a droplet with a V-shaped constellation of 5 point defects in several planes along one of the arms of the V. **a–d** Streamlines of the director field in several planes as indicated in the insets. **e, f** The director field shown with cylinders in the same planes as (**c**) and (**d**), respectively. The scale bars are 5 μm

the exact borders of the cholesteric bubbles are hard to determine, but the bubbles still remain a useful tool for representing the rough outline of the 3D structure of the director field.

At $N \approx 3.2 - 5.4$, another link of a CB with an associated surface $+1$ defect and a hyperbolic -1 defect in the bulk is added to the V-shaped structure to form a string of 7 point defects. Wide field images of two droplets with such a structure are shown in in Fig. 7.37. The appearance of the droplets changes significantly at different focusing depths and at most only a few of the defects can be observed in each plane, which means the string has a 3D structure. The FCPM intensities of the droplet from Fig. 7.37a–c can be seen in Supplementary Movie 1 to Ref. [9]. The positions of defects, deduced from the reconstructed director field are shown in Fig. 7.38. We can see that 5 of the defects lie in a familiar V-shaped arrangement in a non-equatorial plane, but the additional two defects with a CB are expelled from this plane, so that the twisted volume is distributed through-out the droplet. The director field is shown in the plane which includes 5 point defects and intersects 3 CBs, the outline of which

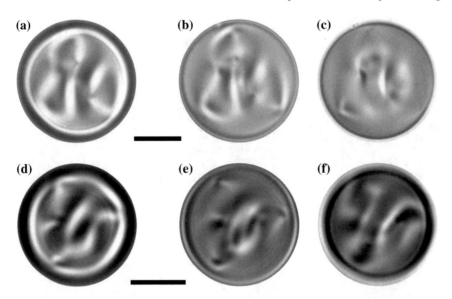

Fig. 7.37 Optical textures of droplets with a string-like constellation of 7 point defects. **a–c** Droplet with 7 point defects, the director field of which is presented Figs. 7.38 and 7.39. The images are focused in different planes: **a** roughly corresponds to a plane with a single +1 point defect in the middle of the string, **b** a plane with two −1 point defects and **c** has an ending arm of the string and a single ending +1 defect. **d–f** Three focusing planes in a differently oriented droplet. The scale bars are 5 μm. Panels **a–c** reprinted from Ref. [9] under the terms of the Creative Commons Attribution 4.0 International License (http://creativecommons.org/licenses/by/4.0/)

can be observed from the streamlines. Similarly as before, each non-terminal CB in the string touches an off-axis hyperbolic point defect with its surface. The two added defects compensate each others topological charge, so the total charge of the droplet stays unchanged. The +1 charges which anchor the CBs in this structure lie approximately in vertices of a tetrahedron as shown in the upper inset in Fig. 7.38.

Figure 7.39 presents the director of the droplet with a string of 7 point defects in more detail. Figure 7.39a–d show the streamlines of the director in a series of slices as indicated in the inset to each slice. The first slice (Fig. 7.39a) shows the cross-section of a cholesteric bubble anchored on one of the non-terminal +1 defects. The streamlines in this cross-section are similar to those from Figs. 7.31b and 7.35a which indicates CBs still retain the cylindrical symmetry they had in the simpler droplets. Figure 7.39b shows the termination of the CB in a plane which includes two −1 point defects similar to Fig. 7.35b and Fig. 7.39c illustrates how the central part of the CB spreads out to form a layer similarly as in the V-shaped structure in Fig. 7.35c but with an extra layer of twist next to the additional arm of the string so that the director rotates by 4π across the diameter. Figure 7.39d shows the streamlines in a plane which goes through a terminal arm of the string and the opposite terminal +1 point defect. The outline of the CB in the arm can be clearly resolved between the ending pair of +1 and −1 defects. Figure 7.39e shows the streamlines in the similar

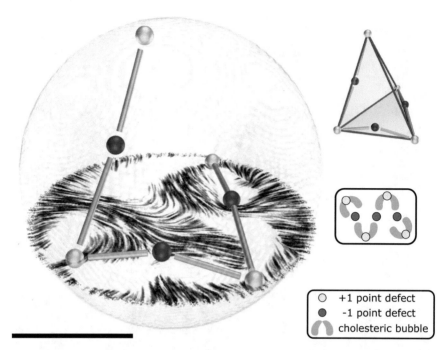

Fig. 7.38 A 3D representation of the droplet from Fig. 7.37a–c. Locations of point defects are indicated with coloured spheres and the cylinders which connect them help to visualise their spatial relation. The streamlines represent the reconstructed director in a plane which includes the lower 5 defects. The +1 defects lie in the vertices of a tetrahedron as presented in the upper inset. The structure is schematically presented in the lower inset. The scale bar is 5 μm. Reprinted from Ref. [9] under the terms of the Creative Commons Attribution 4.0 International License (http://creativecommons.org/licenses/by/4.0/)

corresponding plane going through the other ending arm and the other terminal +1 defect. We can see that the structure is symmetric with regard to the terminal arms of the string—in both planes Fig. 7.39d, e there is an almost identical structure just reoriented accordingly to the point defects.

Figure 7.39f–h illustrate the layeredness of the structure. In Fig. 7.39f we see the streamlines in a plane which separates the droplet into two same, but differently rotated halves. The plane runs through the −1 defect in the middle of the string and through the two +1 defects at the ends of the string. The layers in the middle of the droplet are clearly seen and Fig. 7.39g shows another projection of the structure in a plane that is perpendicular to Fig. 7.39f and goes through both of the −1 defects in the side arms of the string and illustrates how the layers terminate by the director twisting at the edge of the layer to match the homeotropic orientation on the surface of the droplet. Figure 7.39h is also perpendicular to Fig. 7.39f and includes the middle arm of the string, but is positioned in between the two terminal arms of the constellation, along the dark streamlines in the middle of Fig. 7.39f. With this image we can explain how symmetry is achieved between the two terminal arms of the constellation—the

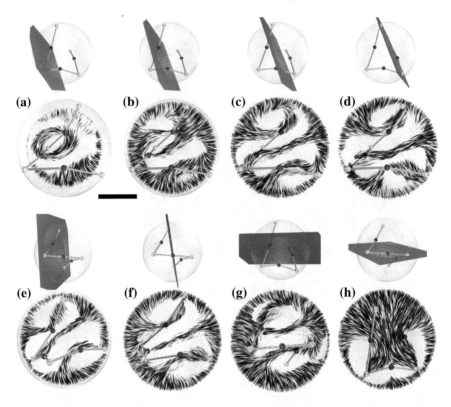

Fig. 7.39 Streamlines of the reconstructed director field of a droplet with a string-like constellation of 7 point defects. The data present the droplet from Fig. 7.37a–c. The positions of the cross-sections are indicated in the insets. Locations of point defects are indicated with coloured spheres and the cylinders which connect them help to visualise their spatial relation. The scale bar is 5 μm

central escapes along the symmetry axis of the two CBs on the $+1$ defects of the central arm of the string merge in a hyperbolic fashion at the middle -1 defect to form a single layer in the bulk of the droplet which connects directly to the surface of the droplet opposite to the middle arm of the string.

Figure 7.39 illustrates that the motif of a cholesteric bubble is still present in these more complex droplets but it's twisting structure merges with cholesteric layers which occupy the bulk of the droplet. Therefore it is hard to determine the exact expanse of a CB and its appearance depends on the plane in which we examine it. Despite this CBs are relatively localised structural elements of the droplets as is illustrated in Fig. 7.40. These two cross-sections dissect the two -1 defects in the side arms of the constellation along their symmetry plane perpendicular to their symmetry axis. We can see that the director field in both these planes is mostly in-plane and almost radial, similar as in Figs. 7.31f and 7.36b. This means that both terminal cholesteric bubbles are relatively separated from the rest of the droplet.

(a)   **(b)**

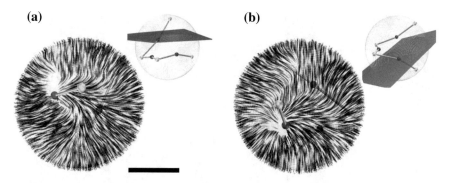

Fig. 7.40 Streamlines of the director field in a symmetry plane of each of the −1 defects in the side arms of the string-like constellation of 7 point defects. The streamlines indicate that the director in these two planes is mostly in-plane and almost radial, similarly as in droplets with strings of fewer defects. The scale bar is 5 μm

At $N \approx 3.9 - 5.2$ strings of 9 point defects running along the surface of the droplet in a spiral fashion can appear in the droplets. Figure 7.41 shows examples optical textures of such droplets in different orientations. Figure 7.41a–c present a droplet with all 9 defects lying around the perimeter of the picture. The series of images was taken at different focusing depths and different structural features of the droplet are stressed in each plane. For example most of the defects can be seen in Fig. 7.41b and in Fig. 7.41a, c the lensing of light on CBs is visible along with point defects which lie outside the central plane. Figure 7.41d shows a polarised image of a similarly oriented droplet under crossed polarisers as indicated in top right corner. The locations of +1 point defects are indicated by pinched areas, with lobes similar to CBs as in the POM images of simpler droplets, but their appearance is somewhat smeared because of the extra twisting in the bigger droplets.

Figure 7.41e–h show a different droplet with 9 point defects, rotated by 90° with regards to the structures in Fig. 7.41a–d, so that one of the non-terminal +1 defects is pointing towards the camera in the middle of the image. The non-polarised image Fig. 7.41e indicates the layers and locations of some of the defects, but the middle +1 is observed only when polarisers are used, for example it is indicated by the central cross when imaged by parallel polarisers in Fig. 7.41f. Areas in which the polarisation rotates by 90° are strongly contrasted in this image and the layers in the bulk of the droplet are clearly visible along with some of the −1 point defects. Under crossed polarisers, the orientation of the polarisers is very important for the visibility of the features, for example in Fig. 7.41g the bright lobes around +1 defects and the dark cross on the middle +1 defect are clearly visible and in Fig. 7.41h with slightly rotated polarisers, their visibility is strongly diminished.

Figure 7.41i–l show a series of non-polarised images at different focusing depths of another slightly differently oriented droplet with the same structure. In this projection the droplet's appearance shows many layers along with some of the point defects, but it would be hard to claim there is any regularity in the structure based just on

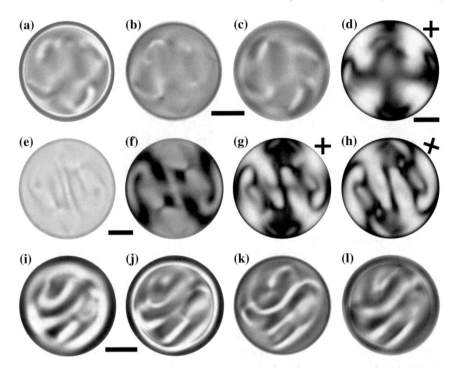

Fig. 7.41 Examples of wide-field optical textures of droplets with strings of 9 point defects. Each row corresponds to a different orientation of the droplet: in the first row the string of defects runs circumferentially around the edge of the droplet and the cholesteric layers in the bulk of the droplet are oriented in the plane of the image. In the second row the layers are perpendicular to the plane of the image and the +1 point defect in the middle of the central V-shaped part is oriented directly towards the viewer. In the third row the layers are tilted out of the xy plane. Image in (**f**) is taken with parallel and those in (**d**), (**g**) and (**h**) with crossed polarisers as indicated in top right corner of each panel. The scale bars are 5 μm. Panels **a–c** reprinted from Ref. [9] under the terms of the Creative Commons Attribution 4.0 International License (http://creativecommons.org/licenses/by/4.0/)

these wide-field images. This series clearly illustrates how wide-field optical textures can be misleading in identifying complex 3D structures of random orientation.

FCPM intensities of the droplet from Fig. 7.41a–c are shown in Supplementary Movie 2 to Ref. [9] and the locations of point defects along with the director field reconstructed from these intensities is shown in Fig. 7.42. We can see how the string of defects spirals along the surface of the droplet. The streamlines represent the director field in a plane running through the central V-shaped part of the string. The schematic representation of the linking of cholesteric bubbles and defects is shown in the inset. All five +1 point defects lie close to the surface of the droplet and stabilise an associated cholesteric bubble and the −1 defects are positioned between the lobes of neighbouring CBs. In this structure the +1 defects do not form a regular polyhedron.

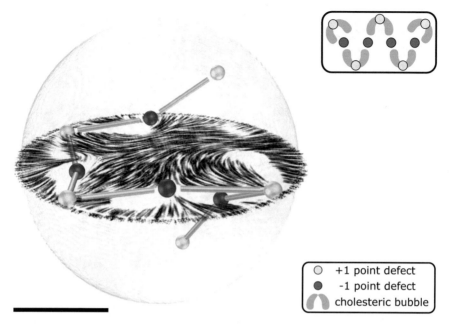

○ +1 point defect
● -1 point defect
⋀ cholesteric bubble

Fig. 7.42 A 3D representation of the droplet from Fig. 7.41a–c. Locations of point defects are indicated with coloured spheres and the cylinders which connect them help to visualise their spatial relation. The streamlines represent the reconstructed director in a plane which includes the middle V-shaped part of the structure. The structure is schematically presented in the inset. The scale bar is 5 μm. Reprinted from Ref. [9] under the terms of the Creative Commons Attribution 4.0 International License (http://creativecommons.org/licenses/by/4.0/)

Figure 7.43 presents the structure with 9 point defects in more detail. Figure 7.43a shows the streamlines in a plane just in front of the two terminal +1 point defects. We can recognise the circular pattern of streamlines which represents the cores of the lobes of the associated two CBs, but we can see the axes of the CBs are not oriented directly towards a neighbouring −1 defect. Figure 7.43b–c illustrate how the two terminal CBs open up into cholesteric layers at the nearest −1 defects to form 5 layers of twist in the bulk of the droplet, as seen in Fig. 7.43d. Figure 7.43e, f show how the layers close and meet the surface at the opposite side of the droplet.

Figure 7.44 is a series of equally spaced parallel slices which present how the director twists in the layers which make up the bulk of the droplet. All the planes are approximately aligned with the cholesteric layers. Figure 7.44a, b include the CB on one of the terminal +1 defect, but in Fig. 7.44c the director is already mostly in-plane. We can see how the direction of the streamlines keeps rotating in the same direction from plane to plane, until reaching the other side of the droplet. In the central portion of the droplet (Fig. 7.44d–j) the cholesteric layer do not reach the surface of the droplet, but are instead separated from it by areas of out-of-plane director which is formed by the CBs. This barrier separates the cholesteric bulk of the droplet from the surface and resolves the mismatch of director orientation.

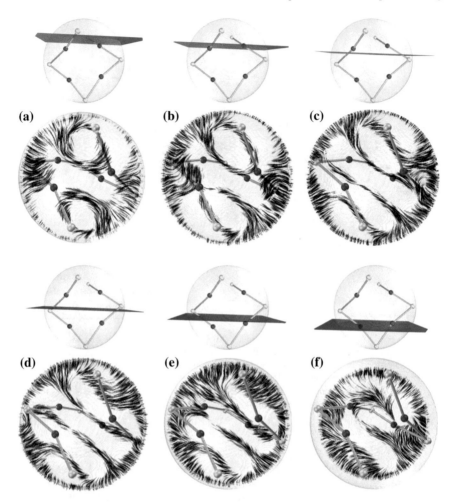

Fig. 7.43 Layerdness of a droplet with a string-like constellation of 9 point defects. The panels show streamlines of the reconstructed director in a series of cross-sections. The data present the droplet from Fig. 7.41a–c. The positions of the cross-sections are indicated in the insets. Locations of point defects are indicated with coloured spheres and the cylinders which connect them help to visualise their spatial relation

This masking of the boundary condition from the layered structure in the bulk of the droplet is similar as the one predicted in numerical quenches in Ref. [6] for the droplet shown in Fig. 3.4. The difference between the two structures is, that in the numerically obtained example, the masking is done with a pair of disclination lines with half-integer winding and here by a series of point defects.

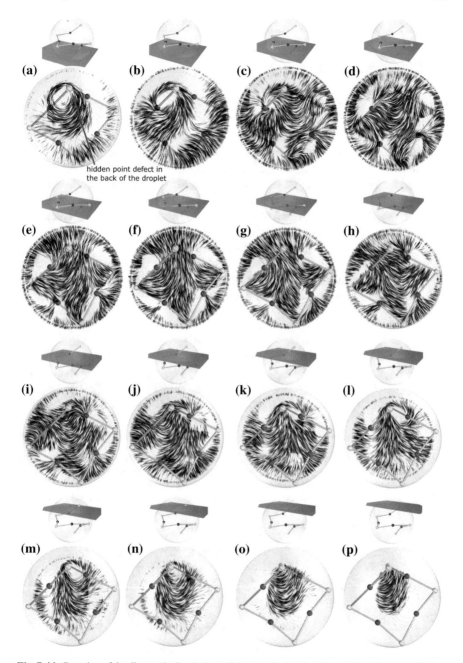

hidden point defect in
the back of the droplet

Fig. 7.44 Rotation of the director in the cholesteric layers of a droplet with a string-like constellation of 9 point defects from Fig. 7.41a–c

7.4   Higher-Charge Topological Point Defects

7.4.1   −2 Point Defect

Some of the textures of the droplets exhibit peculiar symmetry. A structure that appears at $N = 3.7-4.1$ is shown in Fig. 7.45. It is easy to recognise 3 surface point defects indicated by sharp features in Fig. 7.45a, b or by the bright lobes in Fig. 7.45c, d, but the centre of the droplet seems to be occupied with a single structure, hinting that unlike in the previous cases, there is an even number of defects in the droplet. A closer examination of the FCPM intensities in Fig. 7.46 reveals all the singular points of the droplet lie in an equatorial plane. A whole scan of FCPM intensities is shown in Supplementary Movie 3 to Ref. [9]. The I_{tot} in Fig. 7.45e clearly shows a three-fold symmetry of the central part of the droplet.

By examining all the single FCPM polarisations in Fig. 7.46a–d and the I_{tot} in Fig. 7.46e we can see that the discontinuities in the central structure must be concentrated in an area comparable to or smaller than the resolution of the microscope which is around 300 nm. In previous structures we have seen that the minimal distance

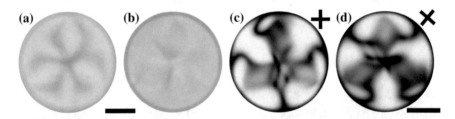

Fig. 7.45 Optical textures of a droplet with a −2 point defect. **a** and **b** are non-polarised microscopy images with different focuses which change the appearance of cholesteric bubbles. **c** and **d** are polarised images with differently oriented crossed polarisers as indicated with a cross in the top right corner of each panel. **c** is the same droplet as in (**a, b**) and (**d**) is a slightly smaller one. The scale bars are 5 μm. Panel a reprinted from Ref. [9] under the terms of the Creative Commons Attribution 4.0 International License (http://creativecommons.org/licenses/by/4.0/)

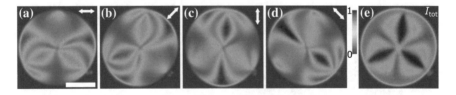

Fig. 7.46 FCPM intensities for a cross-section of a droplet with a central −2 point defect and three +1 surface defects. The panels show FCPM intensities in a equatorial plane in which lie all 4 defects. The polarisation of excitation/detection is marked in the top right corner of each panel. **e** shows the sum of all four polarisations I_{tot} which is proportional to the projection of the director to the xy plane. The scale bar is 5 μm. Panel e reprinted from Ref. [9] under the terms of the Creative Commons Attribution 4.0 International License (http://creativecommons.org/licenses/by/4.0/)

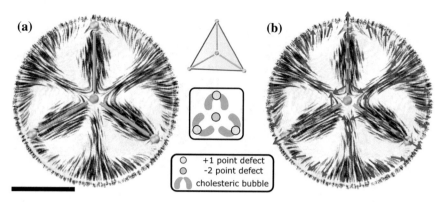

Fig. 7.47 Reconstructed director field of a droplet ($N = 3.7$) with a central -2 point defect in an equatorial plane which includes all the point defects. The FCPM data for this reconstruction are presented in Supplementary Movie 3 to Ref. [9]. Locations of point defects are indicated with coloured spheres and the cylinders which connect them help to visualise their spatial relation. **b** Oriented director field for the plane from (**a**). We can see all the surface defects have a topological charge of $+1$. The scale bar is $5\,\mu$m. Panel a reprinted from Ref. [9] under the terms of the Creative Commons Attribution 4.0 International License (http://creativecommons.org/licenses/by/4.0/)

by which two point defects can be stably separated is on the order of half-pitch of the chiral mixture. In this case that means $3\,\mu$m, which is an order of magnitude larger that the observed size of the defect. We can therefore conclude that the discontinuity in the director field in the centre of the droplet is concentrated in an area of comparable size as an ordinary topological point defect and can regard it as point defect.

The structure of this droplet is shown in Fig. 7.47a with the director in the plane of the defects shown in streamlines, outlining the positions of the CBs as indicated in the schematic representation in the inset. The $+1$ defects lie almost perfectly in the vertices of a regular triangle shown in the upper inset to Fig. 7.47a, not unlike atoms bound to a central trivalent atom, e. g. a carbon atom with sp^2 hybridized orbitals. The reconstructed director field in Fig. 7.47a confirms that the 3 point defects close to the surface are of the same type as in the other droplets, with cylindrical symmetry, the director on the symmetry axis pointing along the axis and twisting away from it, as seen form I_{tot} in Fig. 7.46e. If we assign direction to the director field in Fig. 7.47b, we can see the surface point defects all have $+1$ topological charge. With this it becomes obvious that all defects with radial symmetry which lie close to the surface have $+1$ topological charge almost by definition—a part of their director directly connects with the surface of the droplet, the director on which is by convention oriented outward. This orientation translates to the near-surface radial defect as shown in Fig. 7.47b, giving the defect a positive sign. With this fact we can calculate the topological charge of the central structure: because the total charge of the droplet must be equal to $+1$, the defect in the centre should carry a -2 topological charge: $3 \times +1 + (-2) = +1$. Because this topological charge is concentrated in a region

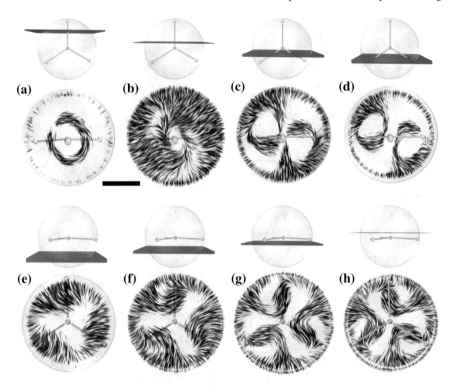

Fig. 7.48 Streamlines of the reconstructed director field of a droplet with a −2 point defect. **a–d** Streamlines in a plane perpendicular to the equatorial plane with point defects. **a** and **b** show two perpendicular cross-sections of one of the CBs which highlight it's cylindrical symmetry and **c, d** show cross-sections of the other two CBs at an oblique angle. **e–h** The director field in planes, parallel to the one in Fig. 7.47. We can see, how the director field rotates across the centre of each cholesteric bubble. The positions of the cross-sections are indicated in the insets. Locations of point defects are indicated with coloured spheres and the cylinders which connect them help to visualise their spatial relation. The scale bar is 5 μm

the size of a point defect we can state that this is a higher-charge topological point defect.

Figure 7.48 illustrates the structure of the droplet with a −2 point defect in more detail. Figure 7.48a shows the familiar cross-section of a CB with a profile of a Bloch skyrmion which makes the CB cylindrically symmetric. Figure 7.48b shows that the lobes of the CB converge toward the central defect as is also seen in the total FCPM intensity in Fig. 7.46e. Figure 7.48c, d illustrate how the director field of the CBs separates into several layers on the other side of the central defect. These cross-sections nicely illustrate the complexity of the twisted structures in the droplets—if a CB is observed in its own reference frame (Fig. 7.48a) it appears a separated entity with well defined cylindrical symmetry, but when observed from a different angle as the other two CBs in Fig. 7.48c, d which are equivalent to the first one because of symmetry it continuously merges with other layers and CBs in the droplet.

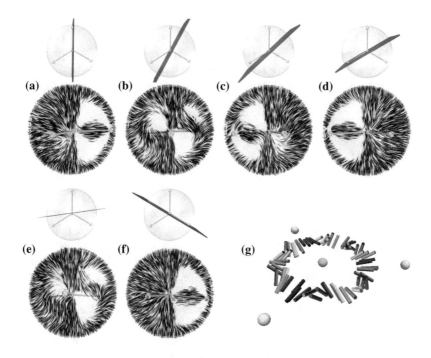

Fig. 7.49 Rotation of the director profile around the central −2 point defect. The director rotates constantly around a direction perpendicular to each of the cross-sections. This behaviour is illustrated by observing the rotation of the director field on a closed loop around the central −2 defect which goes through the centres of all the CBs as shown in (**g**)

To better understand the structure of the −2 point defect, we should examine additional director cross-sections. Figure 7.49 shows the director in a series of planes which include the central −2 defect and are perpendicular to the plane in Fig. 7.47a. The plane in Fig. 7.49a dissects one of the CBs and a surface +1 defect. The director field between the two defects clearly shows the outline of a CB which is anchored on the +1 defect and the director between the CB and the −2 defect has a hyperbolic profile. On the side of the droplet opposite to the +1 defect, the director profile of the −2 defect is radial. Figure 7.49d shows a plane which is rotated by 60° from the plane in Fig. 7.49a around the symmetry axis of the droplet, we can see an identical profile, but the locations are reversed—now the +1 defect with the CB is on the left side and the radial profile is on the left. Figure 7.49b, c show two intermediate planes which illustrate how the director field transitions from the profile in Fig. 7.49a to the one in Fig. 7.49d. We can see that the director field rotates from the hyperbolic to radial director field as we rotate the plane of the cross-section around the symmetry axis of the defect. This transition from radial to hyperbolic director is similar to the one that transforms a radial hedgehog to a hyperbolic one. Figure 7.49e shows that the rotation of the director field in the cross-section continues in subsequent planes and Fig. 7.49f shows the director field in a plane rotated by 120° with regard to Fig. 7.49a. The two director profiles are identical, but the second one is centred on another +1 defect and illustrates that this motif is repeated on each CB. Figure 7.49g shows the rotation of the director field on a closed loop around the −2 defect in the plane that includes all the point defects. We can see that the director on this loop performs a 6π rotation.

7.4.2   −3 Point Defect

At $N \approx 4.3-4.6$ we have also found droplets with similar wide field textures than those with −2 defects in some focusing planes but which showed more variation at other depths as shown in a series of images at different focuses in Fig. 7.50. The three-fold symmetry is visible in all the images but the central part of the image is dark in the part closer to the objective in Fig. 7.50a, b and rotates in the top slices in Fig. 7.50d, e. The centre in the middle part of the droplet in Fig. 7.50c shows additional details compared to Fig. 7.45a, b.

The 3D structure of this droplet, reconstructed from FCPM intensities presented in Supplementary Movie 4 to Ref. [9], is shown in Fig. 7.51. The structure is composed of four +1 defects near the surface, which are located in vertices of a almost regular tetrahedron as indicated in the top inset to Fig. 7.51. Such positioning of the +1 defects is reminiscent of atoms in a molecule with a sp^3 hybridized atom such as carbon, silicon or germanium. The streamlines in the cross-section in Fig. 7.51 which runs through two of the +1 defects and the centre of the droplet clearly show outlines

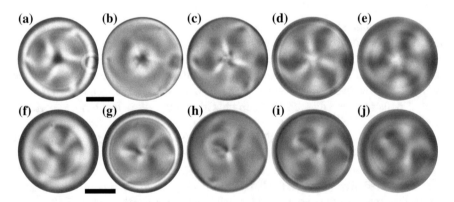

Fig. 7.50 A sequence of non-polarised images of a droplet with a −3 point defect at different focusing depths. The droplet in the top row of images ($N = 4.3$) has one of the symmetry axes aligned with the microscope axis and the one in the bottom row ($N = 4.6$) is slightly tilted compared to the first one. The focus is moved by $3\,\mu$m between subsequent images. Notice the peculiar appearance of the central part of the droplet. The scale bars are $5\,\mu$m. Panels **c** and **e** reprinted from Ref. [9] under the terms of the Creative Commons Attribution 4.0 International License (http://creativecommons.org/licenses/by/4.0/)

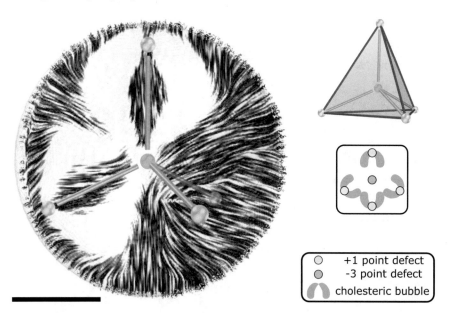

Fig. 7.51 The structure and the reconstructed director field of a droplet ($N = 4.3$) with a central −3 point defect. The FCPM data for this reconstruction are presented in Supplementary Movie 4 to Ref. [9]. Locations of point defects are indicated with coloured spheres and the cylinders which connect them help to visualise their spatial relation. The +1 defects are positioned in vertices of an almost regular tetrahedron as indicated in the inset. The scale bar is 5 μm. Reprinted from Ref. [9] under the terms of the Creative Commons Attribution 4.0 International License (http://creativecommons.org/licenses/by/4.0/)

of cholesteric bubbles which are anchored on the +1 defects in this plane. As we will see later also the other two +1 each have it's own CB as shown schematically in the inset to Fig. 7.51. In the centre of the droplet is a similar singular structure to the −2 point defect, but with a tetrahedral symmetry. We can calculate its topological charge from conservation of topological charge in the droplet—the total charge must be +1 because of the homeotropic anchoring on the surface of the droplet and there are four +1 defects, so the central defect should carry a −3 topological charge.

Figure 7.52 presents the structure of the droplet with a −3 defect in more detail. The streamlines in cross-sections Fig. 7.52a–c which dissect three of the CBs show a three-fold symmetry and are very similar to the ones in a droplet with a −2 defect shown in Fig. 7.48e–h. Unlike in the case of the droplet with a −2 defect, the central defect in this droplet is translated out of the plane of the three +1 defects to achieve tetrahedral symmetry. Figure 7.52d shows how the cholesteric bubbles merge and close at the central defect which is positioned in an equatorial plane and the almost in-plane director field in Fig. 7.52e in a plane just above the −3 defect signifies that the CBs are relatively separated structures, similarly as in other droplets. Figure 7.52f, g show the typical cross-section profile of a CB and Fig. 7.52h the twisted director profile in the mid-plane of the top +1 point defect.

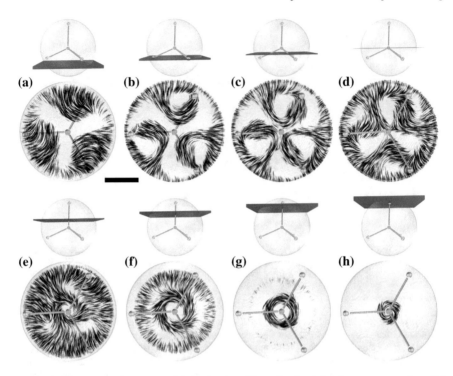

Fig. 7.52 Reconstructed director field of a droplet with a -3 point defect in a sequence of parallel planes along one of the symmetry axes of the droplet running through the -3 defect and one of the $+1$ defects. The part of the droplet shown in (**a–c**) is very similar to the structure in a droplet with a -2 defect. Panel **d** shows how the CBs of the bottom defects close up and in (**e**) the director is mostly in-plane, separating the top bubble from the rest of the droplet. Panels (**f, g**) reveal a typical profile of a CB and (**h**) shows the twisted director field in the middle plane of the top $+1$ defect. The positions of the cross-sections are indicated in the insets. Locations of point defects are shown with coloured spheres and the cylinders which connect them help to visualise their spatial relation. The scale bar is $5\,\mu m$

Figure 7.53 further illustrates the symmetry of the droplet in a sequence of cross-sections rotated by $15°$ around a symmetry axis going through one of the $+1$ defects and the central -3 defect. Slices Fig. 7.53a, e, i are separated by $60°$ and each dissects a cholesteric bubble. Slices Fig. 7.53a and i are nearly identical, but slice Fig. 7.53e is displaced from the location of the $+1$ defect by a small angle, which could also be due to experimental error in determining the location of the defect. The director in this slice is very similar to the other two, but it is their mirror image as it dissects a CB on the other side of the droplet. In fact the whole motif of rotation of the director field around a direction normal to the cross-section in the bottom part of the droplet in slices Fig. 7.53b–d repeats itself exactly in Fig. 7.53j–l and in its mirror image in Fig. 7.53f–h. This bottom part of the droplet with three-fold symmetry is very similar to the structure in the droplet with a -2 defect with the difference that the $+1$ defects don't lay in an equatorial plane and that the -3 defect is translated out of the plane

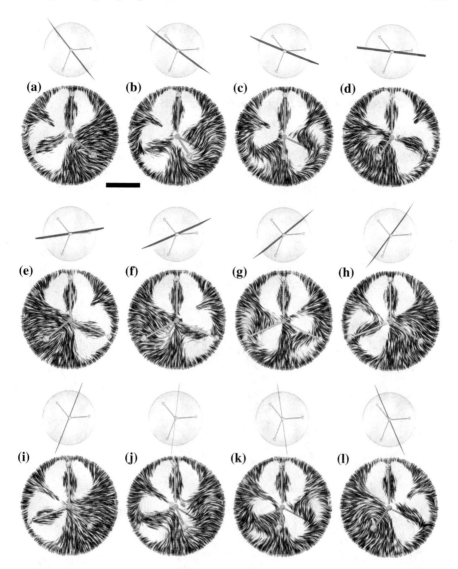

Fig. 7.53 Rotation of the director profile around the central −3 point defect. The plane of the cross-section is rotated by 15° around the vertical symmetry axis. In the bottom part of the droplet, the director rotates constantly around a direction perpendicular to each of the cross-sections, similarly as in the droplet with a −2 defect. In the top part of the droplet above the −3 defect, the director profile does not change because of the cylindrical symmetry of the cholesteric bubble. The scale bar is 5 μm

of the +1 defects towards the centre of the droplet. The part of the droplet above the −3 defect has a constant profile in all slices in Fig. 7.53 because of the cylindrical symmetry of the CB which is centred on the +1 defect above the central defect.

All the CBs are equivalent, so the structure can be understood as 4 symmetrically positioned CB around a central 4-valent point defect in a tetrahedral structure.

7.4.3 Structure of the Higher-Charge Point Defects

There is an easier way to measure the topological charge of a defect than to explicitly calculate the integral in Eq. (2.17). The first step is again to surround the structure with a sphere and prescribe orientation to the director on this sphere as seen in Fig. 7.54a, b. A derivation from Ref. [10] instructs us to decompose the director field on a surface into different building blocks, from which we can calculate the topological charge. If no "grains" are present, we only need to count the number of patches with inverted director direction on an aligned background. The enclosed topological charge is then calculated as $q = 1 - M$ where M is the number of patches on the surface. The background, inwards pointing director in Fig. 7.54a, b is coloured blue and the outward pointing patches are red. Please note, that we have inverted the direction of

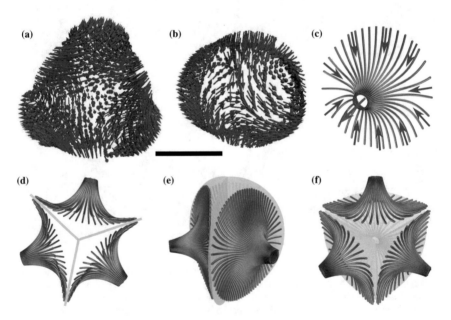

Fig. 7.54 Structure of the higher-charge defects. **a, b** Two different views of the experimental director field with arrows around a -2 point defect. The colours show if the director is pointing out (red) or in (blue) on the surface. The scale bar is $2\,\mu$m. The direction of director is inverted compared to other calculations in this Thesis to make the patches more visible. **c** A schematic representation of a hyperbolic pathc. **d** Top and **e** side view of a model structure of a -2 point defect. **f** The model structure of a -3 point defect. One of the patches is pointing away from the viewer. Reprinted from Ref. [9] under the terms of the Creative Commons Attribution 4.0 International License (http://creativecommons.org/licenses/by/4.0/)

director compared with previous calculations to make Fig. 7.54a, b clearer. As the number of patches on the surface is $M = 3$, it follows that the enclosed topological charge is -2. This result agrees with the previous calculation of the total topological charge in the droplet.

A way to understand the structure of the -2 defect is to first examine an ordinary hyperbolic defect with unit charge. If we assume that the vectorised director in the radial plane of the defect is pointing toward the defect, the hyperbolic profile along the symmetry axis of the defect forms two patches of outgoing director. This gives the defect a $+1$ topological charge, because we inverted the director arrows compared to the usual choice. A single hyperbolic patch from this defect is shown in Fig. 7.54c. We can see that a patch is an area where the director lines converge and lead away from the point defect because of its hyperbolic structure. A model of a -2 defect can be constructed, if we insert another patch into the structure of a hyperbolic unit-charge defect. The radial part of each patch now cannot span a whole $180°$, but instead must be contained in a $120°$ angle. This gives rise to the three-fold symmetry of the defect which can be seen in the schematic structure shown in Fig. 7.54d, e.

We can build a model of the -3 defect in a similar way. Because it's charge is -3, it must have $M = 4$ patches. We can start with the -2 defect in Fig. 7.54d, which has three-fold rotational symmetry and try to add one more hyperbolic patch. We could add it in the plane of the other three patches, but if we insert it from a perpendicular angle, the volume available to the cholesteric bubbles which surround the defect is maximised. By inserting the fourth patch, we distort the structure of the other three, tilting them out of the plane to form a tetrahedral structure as shown in Fig. 7.54f.

We can use these patches to understand the structure of the droplets with higher-charge defects. Each patch of a defect is an attachment point for a neighbouring $+1$ defect with an associated cholesteric bubble. In the case of a -1 defect there are two patches and it is surrounded by two $+1$ defects as for example in a droplet with 3 collinear point defects. A -2 point defect has 3 patches and each patch points towards a neighbouring $+1$ point defect close to the surface of the droplet. The -3 defect has 4 patches, and each of the patches is oriented towards a neighbouring $+1$ point defect with an associated CB as shown in Fig. 7.53a, e, i. Symmetrical tetrahedral positioning of the $+1$ defects maximises the volume available to each CB and therefore minimises the elastic distortions of the director field.

The symmetry of structures with the higher-charge point defects evokes an analogy with material molecules that include atoms, which form bonds with more than a single neighbouring atom. A simple -1 hyperbolic point defect with two patches can be positioned between two pairs of a $+1$ point defect and a CB, as in the droplet with a simple string of 3 collinear point defects with two CBs. A chemical analogue would be a carbon dioxide molecule, where the two oxygen atoms are positioned on opposite sides of a carbon atom to form a linear structure. The trifold symmetry of the -2 defect and the planar arrangement of the surrounding $+1$ point defects is reminiscent of a molecule with a carbon atom with three sp^2 hybridised orbitals, which are positioned in a triangular structure and can bond with three surrounding atoms. This makes the carbon atom and in a topological sense our -2 point defect three-valent. Similarly the -3 point defect can be compared with a carbon atom with sp^3 hybridised orbitals. These 4 orbitals are arranged in a tetrahedral structure and allow chemical bonds with 4 surrounding atoms, just like in the case of a droplet with a -3 defect and four $+1$ defects. We can therefore say that the -3 topological point defect has a valency equal to 4.

There is one important difference between the schematic model of the higher-charge point defects we have constructed here and the observed structure of the defects, shown in Fig. 7.54a, b: the predominant elastic deformation in the hyperbolic model is bend, and in the reconstructed data it can be clearly seen that the director changes orientation between the patches and the background by twisting. The two structures are topologically equivalent, because they can be transformed between each other with a rotation of the director field in a similar way as a radial point defect can be transformed into a hyperbolic one. The difference between the two is therefore in the energetics of the deformation and not in the topology, so the hyperbolic patch structure can nevertheless be used to explain the structure of the higher-charge defects.

An open question about the higher-charge point defects is the exact nature of their defect cores: in principle they could be true point defects, or their cores could be formed by defect loops in an analogous fashion as the recently demonstrated cores of the $+1$ point defects [11], but with a complex arrangement of the director in their cross-sections, which would allow them to carry non-unit topological charge [5]. Another possibility for the core of the higher-charge defects is that they would be composed of several tightly packed unit charge point defects. For example, the -2 point defect which has a three-fold rotational symmetry would have to be composed of three -1 point defects, positioned in vertices of a triangle with an additional $+1$ point defect in the centre of the triangle to achieve a total topological charge equal to -2. However, the FCPM images of the -2 defect show, that the core of the defect is confined to an area of similar size as the other point defects. Typical distances over which the helical twisting of the director can stabilise the point defects of opposite topological charge against annihilation are of the order of half of the cholesteric pitch. In the proposed composite structure of the -2 defect core, the distance between the ± 1 defects is much smaller than the pitch, which means that the defects would have to be stabilised by a different mechanism for this structure to be feasible. A similar argument could be made for the core structure of the -3 point defect. The resolution of our FCPM method is around 300 nm, and because the core structure is confined to this volume, the logical conclusion is that the higher-charge defects are point defects in the same sense as the unit-charge ones.

7.5 Topological Molecules

7.5.1 Topological Molecules with Higher-Charge Defects

In some droplets parts of previously presented structures are substituted by more complex substructures. For example Fig. 7.55 shows wide-field images of a droplet ($N = 3.9$) with a structure similar to the droplet with a -2 defect in the centre and three $+1$ point defects close to the surface, where one of the $+1$ defects is substituted with a chain of three collinear point defects, similar to the one found in a droplet with a string of 3 defects presented in Sect. 7.3. Such a structure appears at $N \approx 3.9 - 4.5$.

A 3D representation of the structure with a streamline representation of the director in one plane is shown in Fig. 7.56. The similarity with the simpler droplet with a -2 defect is obvious from the streamlines presented in this figure which reveal the outlines of the CB around the -2 defect. One of the CBs appears deformed—it is anchored on a chain, formed of two $+1$ defects close to the droplet surface and a -1 defect centred between them and the chain is oriented perpendicularly to the plane of the other defects. The $+1$ defects are positioned close to the surface of the droplet and as far apart as possible, giving the whole structure tetrahedral symmetry instead of the triangular of the simpler droplet with a -2 defect. The positioning of the $+1$ defects in vertices of a regular tetrahedron is shown in the top inset to Fig. 7.56 and the structure is schematically presented in the bottom inset.

More details about the structure are shown in Fig. 7.57. Figure 7.57a, b present cross-sections which include the -2 defect and one of the $+1$ defects next to it. We can see the outlines of CBs between the two defects, but on the other side the director profile is almost radial but slightly twisted, similarly as in the corresponding cross-sections of the simpler structure with a -2 point defect in Fig. 7.49a, d, f.

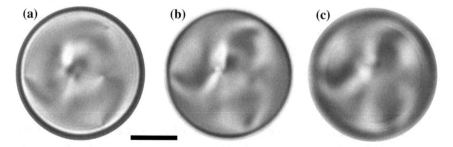

Fig. 7.55 A sequence of non-polarised images of a droplet ($N = 3.9$) with a -2 point defect with two $+1$ point defect plus a string of three point defects. The images are taken at different focusing depths separated by $3\,\mu$m starting from the objective. The string of 3 defects is oriented vertically in the plane of the panel (**a**), where another $+1$ defect is located on the left side of the droplet. In panel (**b**) the -2 defect is located just up and left of the centre, and in (**c**) the remaining $+1$ defect is a bit up from the centre. The scale bar is $5\,\mu$m. Reprinted from Ref. [9] under the terms of the Creative Commons Attribution 4.0 International License (http://creativecommons.org/licenses/by/4.0/)

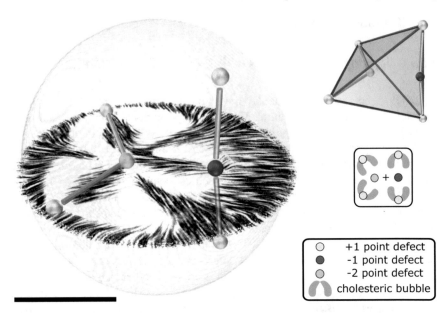

○   +1 point defect
●   -1 point defect
◯   -2 point defect
⌒   cholesteric bubble

Fig. 7.56 The structure and the reconstructed director field of a droplet ($N = 3.9$) with a -2 point defect, where one of the $+1$ defects from Fig. 7.47 is substituted with a string of 3 collinear point defects. The FCPM data for this reconstruction are presented in Supplementary Movie 5 to Ref. [9]. Locations of point defects are indicated with coloured spheres and the cylinders which connect them help to visualise their spatial relation. The $+1$ defects are positioned in vertices of an almost regular tetrahedron as indicated in the inset. The streamlines are shown for a plane going through the -2 defect, its two surrounding $+1$ defects and the -1 defects from the string of 3 defects. The structure is schematically shown in the bottom inset. The scale bar is $5\,\mu$m. Reprinted from Ref. [9] under the terms of the Creative Commons Attribution 4.0 International License (http://creativecommons.org/licenses/by/4.0/)

Figure 7.57 shows that the CBs between the -2 defect and the two neighbouring $+1$ defects is cylindrically symmetric and has a skyrmionic cross-section. Figure 7.57d is the cross-section which includes the -2 defect and the string of collinear defects. We can see that what appeared as a CB in Fig. 7.56 are actually lobes of two different CBs anchored on both $+1$ point defects of the string. These two CBs are actually delocalised and merge with cholesteric layers as shown in Fig. 7.57e. Figure 7.57f shows streamlines in a plane a bit above the plane of the -2 defect and its two associated $+1$ defects. The profile is very similar to Fig. 7.48g in the simpler droplet with a -2 defect, but the -2 defect is translated out of centre and the space which opens with this move is filled with the string of collinear defects. Figure 7.57g–l present a series of parallel slices. Figure 7.57g shows that the two CBs next to the -2 defect are slightly asymmetric and Fig. 7.57h, i illustrate how this asymmetry transforms into parallel cholesteric layers in Fig. 7.57j. Figure 7.57k, l show how the layers merge close to the string of defects to form the two asymmetric CBs in the string.

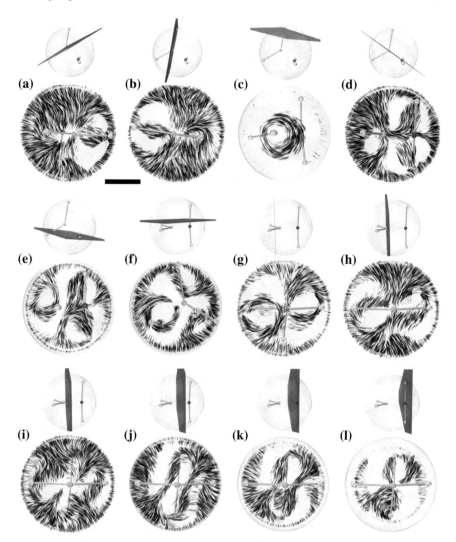

Fig. 7.57 Reconstructed director field of a droplet with a −2 point defect where one of the +1 defects is substituted with a string of 3 collinear defects. Streamlines are shown for various cross-sections as shown in the insets. **a, b** show the cross-sections through the −2 defect, which include one of the connected +1 defects. Outline of the CB in each slice is visible along with the almost radial but slightly twisted configuration on the other side. **c** shows the typical cross-section of one of the CBs next to the -2 defect. **d** is a slice which includes the −2 and the string of 3 defects and **e** is slightly rotated around the string, so it dissects a CB next to the −2 defect. **f** is a plane above the plane of the −2 structure and **g–l** show a series of parallel slices

Another example of a composite structure with a −2 defect which appears at $N \approx$ 4.2−4.5 is shown in Fig. 7.58. The transmission illumination images in Fig. 7.58a–e appear to be very similar to the ones of a droplet with a −3 defect, but the details

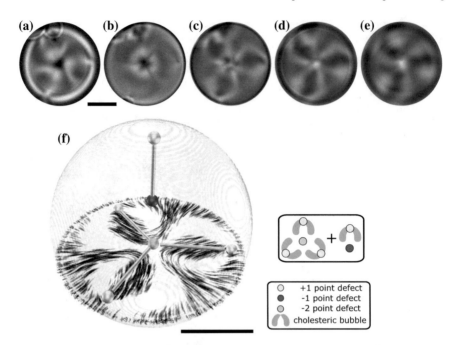

Fig. 7.58 A droplet ($N = 4.2$) with three collinear defects, where one of the $+1$ defects is substituted with a triangular structure made of a central -2 defect and three $+1$ defects. **a–e** A sequence of non-polarised images at different focusing planes separated by $3\,\mu$m, **a** being the closest to the objective. The symmetry axis of the droplet is oriented perpendicularly to the plane of the images. **f** The 3D structure of the droplet with streamlines in the plane of the bottom 3 defects. The -2 defect is slightly above this plane and lies on the symmetry axis of the droplet together with the -1 defect and the top $+1$ defect. The relative positions of the point defects and the cholesteric bubbles are shown schematically in the inset. The scale bar is $5\,\mu$m

in the central part of the image in Fig. 7.58c, d appear to be a bit different as shown in the comparison. The reconstruction performed from experimental FCPM data (Fig. 7.58f) explains this by showing that the structure has an additional point defect on its symmetry axis. This structure is analogous to a string of 3 collinear defects, where one of the $+1$ defects is substituted with the whole structure of a droplet with a -2 defect and three $+1$ defects. The whole substitute structure is carrying a total topological charge of $+1$ just as the replaced point defect does. The substituted portion of the droplet forms a equilateral triangle with the three $+1$ point defects in its vertices and the streamlines in this plane shown in Fig. 7.58f reveal the outlines of three CBs, each anchored on one of the $+1$ defects and oriented toward the -2 point defect. A pair of ±1 defects which is separated with another CB is positioned normal to the triangle on the symmetry axis of the droplet. Because of the repulsion between the -1 and -2 defects and the cholesteric bubble centred on the top $+1$ defect, the tetrahedron formed by the four surface $+1$ defects is elongated along the symmetry axis of the structure.

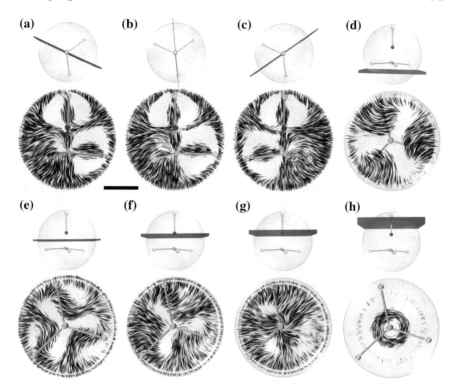

Fig. 7.59 Director streamlines in a selection of cross-sections for the droplet in Fig. 7.58. Cross-sections in (**a–c**) include the symmetry axis of the droplet and one of the bottom +1 point defects and show the outlines of cholesteric bubbles. Panels **d–f** show the streamlines in the bottom part of the droplet which is nearly identical to the simplest droplet with a −2 defect, shown in Fig. 7.48e–h. **g** The cross-section through the −1 point defect, where the director profile is similar as in Fig. 7.52e which is just above a −3 defect. **h** The typical cross-section of a cholesteric bubble. The scale bar is 5 μm

Figure 7.59 presents additional cross-sections of the structure. Figure 7.59a–c show cross-sections which include the symmetry axis of the droplet and one of the bottom three +1 point defects as shown in the insets. We can see that each of the bottom +1 defects has a CB anchored on it, orient towards the −2 defect, and that the top CB is cylindrically symmetric with slight modulation of its bottom outline because of proximity of the other CBs. Slices Fig. 7.59d, e together with the streamlines Fig. 7.58f show that the bottom part of the droplet very closely corresponds to the structure in Fig. 7.47, with the −2 defect translated just slightly out of the plane of the three surrounding +1 defects towards the centre along the symmetry axis of the droplet. Figure 7.59f, g show how unlike Fig. 7.48 the vertically oriented director in the central three-fold cross reorients in-plane to accommodate the hyperbolic −1 point defect on which the top ±1 pair is based. Figure 7.59h reveals the typical cylindrically symmetric skyrmionic profile of the top CB.

This droplet is very similar to the droplet with a -3 defect and four $+1$ defects, the biggest difference being that the negative charge in the bulk of the droplet is not concentrated at a single region but in two point defect, which are separated by a distance on the order of half-pitch of the CLC mixture. This separation can be clearly seen in Fig. 7.59a–c which differ from Fig. 7.53a, e, i—in the latter the bottom cholesteric bubbles are oriented towards the centre of the droplet whereas in the former their orientation is much more in-plane towards the lower-positioned -2 defect. A difference can also be clearly seen in slices at and just above the two higher-charge defects in Fig. 7.59e where the structure is opened up similarly as in the simpler structure with a -2 defect in Fig. 7.48h, to transition from the structure of the three in-plane CBs to the almost radially oriented director above it and then close up again in Fig. 7.59f, g to accommodate another CB similarly as in Fig. 7.52d, e immediately above the -3 defect.

Composite structures can also include -3 defects as in the example at $N = 4.7$, shown in Fig. 7.60. Locations of point defects and outlines of the CBs can be seen in the non-polarised transmission images Fig. 7.60a–d. Figure 7.60e shows the streamlines of the director field in an equatorial plane at the same orientation of the droplet as in Fig. 7.60a–d. We can see a close correlation between the outlines of CBs in the transmission images and the locations of the bubbles in the streamlines. Figure 7.60f shows the structure in another orientation with the streamlines in the droplets mirror symmetry plane.

The 3D structure of the droplet can be understood as a substitution of one of the $+1$ defects in a droplet with one -3 and four $+1$ defects with a V-shaped string of 5 charge alternating defects. All the $+1$ defects in the structure are positioned close to the surface of the droplet, but because their number is increased from 4 to 6, the symmetry of the structure changes from tetra- to octahedral as indicated in the top inset with a geometric solid. In the streamlines in Fig. 7.60f we can recognise the locations of two CBs which the plane dissects. One is located between one of the $+1$ defects in the bottom of the droplet and the -3 defect and the other one is between the -3 defect and the V-shaped constellation in the top of the droplet. This bubble is deformed compared to the central bubble in a V-shaped constellation from Fig. 7.34c, d, with its lobes extending past the two -1 defects and merging with the cholesteric layers which fill the area between the V and the -3 defect.

Figure 7.61a–d shows the streamlines in a series of horizontal slices. In Fig. 7.61a we can see that the structure just below the V-shaped constellation has a similar profile to that of the simpler V-shaped constellation in Fig. 7.34a, b. Slices further down toward the -3 defect show how the director rotates in the central part of the droplet and the CBs finish above the -3 defect—most of the director in Fig. 7.61c is in-plane, except directly above the -3 defect. Figure 7.61d shows the three fold rotational symmetry of the director around the -3 defect, dissecting the bottom three CBs. Figure 7.61e–h shows another series of parallel slices oriented as indicated in the insets. Figure 7.61e shows the typical cylindrical symmetry of three CBs and how the regions which separate them extend over the whole droplet. Figure 7.61f illustrates the termination of the central CB of the V-shaped constellation at the two -1 defects, similarly to the structures in droplets with strings of defects. Figure 7.61g,

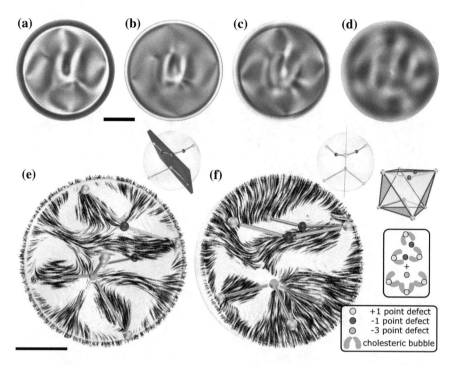

Fig. 7.60 A droplet ($N = 4.7$) with a -3 defect, where one of the $+1$ defects is substituted with a V-shaped constellation of 5 defects. **a–d** A sequence of non-polarised images at different focusing planes separated by $3\,\mu$m, **a** being the closest to the objective. **e** Streamlines in an equatorial plane of the droplet with an orientation that roughly corresponds to the transmission images in (**a–d**). **f** Another projection of the streamlines in a plane with most symmetry. This plane includes the -3 defect and one of the $+1$ defects around it and the $+1$ defect in the middle of the V-shaped constellation. The octahedral symmetry of the positions of $+1$ defects is highlighted in the top inset to (**f**) and the relative positions of the point defects and the cholesteric bubbles are shown schematically in the bottom inset. The scale bars are $5\,\mu$m. Panels b and f reprinted from Ref. [9] under the terms of the Creative Commons Attribution 4.0 International License (http://creativecommons.org/licenses/by/4.0/)

h show cross-sections which are analogous to Fig. 7.61e but on the other side of the -3 defect. We can resolve the outlines of the CBs and they are separated from each other in a similar way as in Fig. 7.61e.

Figure 7.62a–f present another projection of the droplet in a series of parallel slices. Figure 7.62a shows the cylindrical symmetry of one of the CBs next to the -3 defect, and Fig. 7.62b shows the two CBs in one of the arms of the V-shaped constellation, with one of the lobes extending toward the -3 defect. Figure 7.62c goes through the -3 defect and shows the complex director patterns around it with a lot of out of plane twisting. We can see that the extended lobes of the top CBs connect to the -3 defect. In Fig. 7.62d we can see two CBs on two of the bottom $+1$ point defects and the layered structure in the top part of the droplet, which closes up

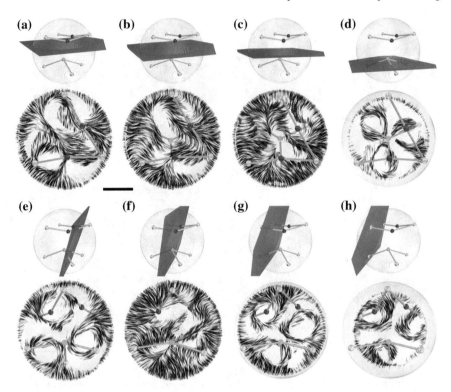

Fig. 7.61 Octahedral droplet with a −3 defect. **a–d** director streamlines in a series of parallel cross-sections. **a–c** show the director structure between the V-shaped constellation and the bottom part with the −3 defect. **d** shows oblique cross-sections of the three cholesteric bubbles around the −3 defect. **e–h** Streamlines in a series of cross-sections: **e** dissects three CBs as shown in the inset above the pictures, **f** is a plane which includes all the negative charges, **g** dissects the other three CBs and **h** is a plane with three +1 defects. The scale bar is 5 μm

at the other −1 defect in Fig. 7.62e to form another CB. The cross-section of this CB can be seen in Fig. 7.62f and it is separated from the rest of the droplet by the out of plane director in Fig. 7.62e, f. Figure 7.62h, g show cross-sections of two more CBs, both having cylindrical symmetry and being separated from the rest of the droplet in a similar way as the one in Fig. 7.62f.

All the presented topological molecules can be understood as permutations of stable structures presented earlier, where one of the surface +1 point defects is substitute with a larger structure with an equivalent topological charge. In the presented examples, a point defect is substituted with a linear string of 3 point defects, a V-shaped string of 5 point defects or a triangular constellation of three +1 defects arranged around a central −2 point defect. All three substitutes are either linear or at most planar, as a 3D arrangement of defects in the replacement structure would probably take up too much volume in the droplet.

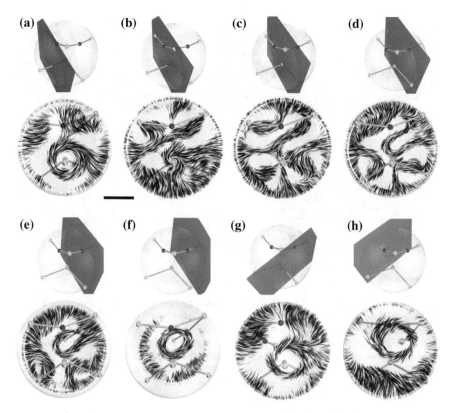

Fig. 7.62 Director streamlines in a selection of cross-sections for the octahedral droplet with a -3 defect. **a** Cross-section of a CB on one of the $+1$ defects around the -3 defect. **b** A cross-section which includes the three defects in one of the arms of the V-shaped constellation. **c** The director structure in the bulk of the droplet. This plane includes the -3 defect. **d** A plane which includes two of the bottom $+1$ defects and dissects the middle top CB. **e** A similar plane as (**d**), but tilted, so it includes one of the -1 defects. **f–h** Cross-sections which illustrate the typical cylindrically symmetric profiles of three cholesteric bubbles. The scale bar is $5\,\mu m$

7.5.2 Topological Molecules with Disclination Lines

Substitutions of point defects are also possible in droplets with line disclinations. Figure 7.63 shows a droplet in which a ring disclination runs around two double twist cylinders similarly as in Fig. 7.21 but where one of the point defects terminating a cylinder is substituted with a string of 3 charge alternating point defects. Figure 7.63a–e show non-polarised transmission images of the droplet and the disclination line can be seen in the front part of the droplet in Fig. 7.63a, b and at the back in Fig. 7.63c, d. The string of three point defects can be seen in lower right part of Fig. 7.63c, d and the connected cylinder extends from them towards left. The other double twist

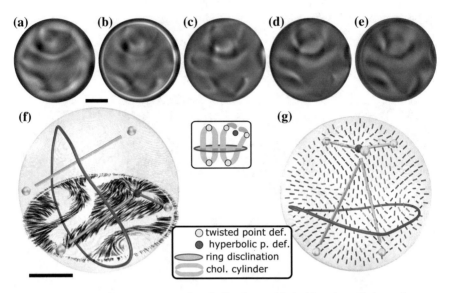

Fig. 7.63 A droplet ($N = 6.4$) with a ring disclination and 2 double twist cylinders, where one of the defects which terminate a cholesteric cylinder is substituted with a string of 3 defects. **a–e** A sequence of non-polarised images at different focusing planes separated by 3 μm, **a** being the closest to the objective. **f** 3D representation of the structure with locations of point defects marked by coloured points. The long rods schematically present the positions of double twist cylinders. Streamlines show the director field in a plane going through the cylinder with the string of defects. The full director field of this plane is shown in (**g**). The scale bars are 5 μm

cylinder lies at an oblique angle and its beginning can be seen in the top part of the pictures.

Figure 7.63f shows the 3D structure of this droplet with the long rods indicating the axes of the two double twist cylinders. The streamlines show the director field in a plane which dissects the cylinder with the string of defects and Fig. 7.63g shows the full director field in this plane. We can see, that the director rotates continuously except in the flattened part of the cylinder next to the string of defects, where the structure is more layer-like. The director around the disclination line has a profile of a twist disclination and it is separated from the central cylinders by π twist.

Figure 7.64 presents the structure in more detail. First, it shows a series of parallel slices in Fig. 7.64a–e on which we can track the ring disclination. In the first slice in Fig. 7.64a the location of the disclination is indicated by the director which is rotated by $\pi/2$ with regard to the anchoring direction and is pointing along the disclination. The whirlpool-like director profile in this slice indicates the location of one of the defects which terminate the cholesteric cylinders. In the next slice the crossing of the disclination is indicated with either a $+1/2$ or a twisted profile. Figure 7.64b dissects the simpler cylinder in the middle at an oblique angle and Fig. 7.64c includes the other point defect of the simpler cylinder. In Fig. 7.64e the disclination is seen at the edge where the director is out of plane. Figure 7.64f, g show two cross-sections

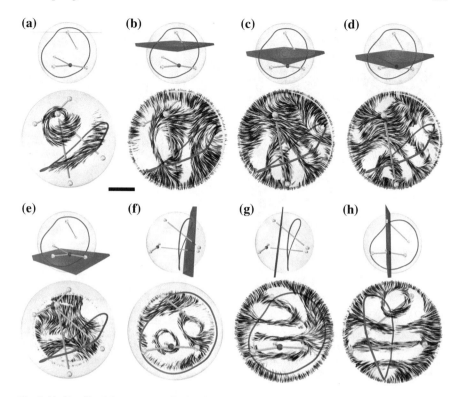

Fig. 7.64 Details of the structure of a droplet ($N = 6.4$) with a ring disclination and 2 double twist cylinders, where one of the defects which terminate a cholesteric cylinder is substituted with a string of 3 defects. **a–e** The director streamlines in a series of parallel cross-sections. **f** Cross-section of the two cylinders close to the ring disclination. **g** Another cross-section of the two cylinders, showing how their profile flattens closer to the string of defects. **h** Streamlines in a plane along the cylinder with the string of defects. The scale bar is 5 μm

of the cylinders, the first one close to the disclination line where the cylinders are narrow and symmetric and the second one closer to the string of 3 defects where the cylinders are flattened to an almost layer-like structure. Figure 7.64h shows a cross-section along the middle of the cylinder with the string of 3 defects, illustrating that despite the substitution of one of the defects with a more complex formation the longitudinal profile of a cylinder stays very similar to the ones in simpler droplets.

Ring disclinations can appear also in combination with cholesteric bubbles. An example of that is shown in Fig. 7.65. In the transmission images Fig. 7.65a–e we can see the disclination line in the top part of the droplet in the planes closer to the microscope objective (Fig. 7.65a, b) and in the bottom part of the droplet in the deeper focusing planes (Fig. 7.65d, e). In Fig. 7.65b, c we can identify four point defects which are nearly collinear on a vertical line and between pairs of defects we can recognise the outlines of the lobes of two cholesteric bubbles. A 3D representation of the structure is shown in Fig. 7.65f, where the streamlines in the equatorial plane

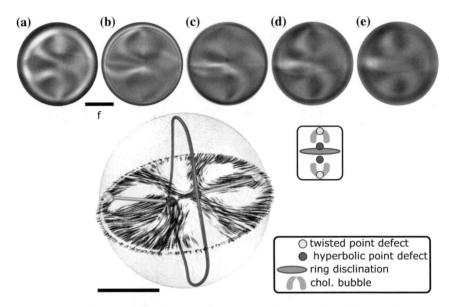

Fig. 7.65 A droplet ($N = 4.9$) with a ring disclination and 2 cholesteric bubbles. **a–e** A sequence of non-polarised images at different focusing planes separated by $3\,\mu$m, **a** being the closest to the objective. **f** 3D representation of the structure with locations of point defects marked by coloured points. Streamlines show the director field in an equatorial which is approximately perpendicular to disclination and includes three of the point defects. The scale bars are $5\,\mu$m

which includes three of the point defects indicate the positions of the CBs. We can see that the fourth point defect lies off-axis compared to the other defects. A ring disclination runs circumferentially around the droplet approximately in a plane which is perpendicular to the plane with streamlines.

Figure 7.66a–g presents director streamlines in a series of parallel slices. In the centre of Fig. 7.66a we can see a region of director which is oriented perpendicular to the surface of the droplet (horizontal orientation in the image). The disclination line runs between this region and the surface of the droplet, and locations where the disclination pierces the plane are indicated by white, out-of-plane regions. in Fig. 7.66b, c we can see the edges of the lobes of the two CBs between each pair of defects and in Fig. 7.66d we can see the outlines of the CBs and the twisted region which separates them. Figure 7.66e–g are approximate mirror images of Fig. 7.66a–c and Fig. 7.66h is the plane which separates the two CBs. The director in the centre of this plane is oriented almost azimuthally but Fig. 7.66b, c, e, f show that it is tilted out-of-plane of the figure Fig. 7.66h.

The structure of the droplet is shown schematically in the inset to Fig. 7.65. The structure is in a way similar to the one of a droplet with 2 CBs and 3 collinear point defects, but with the director twisting by 2π in the volume between the CBs instead of

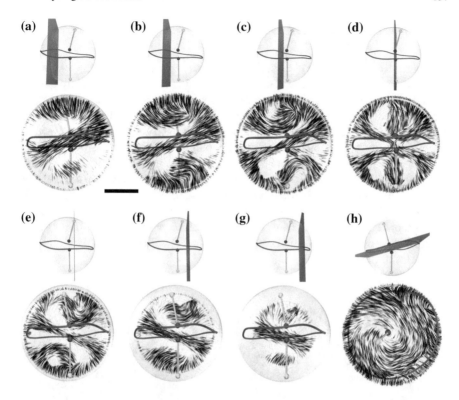

Fig. 7.66 Details of the structure of a droplet from Fig. 7.65. **a–g** The director streamlines in a series of parallel cross-sections, **d** includes three of the point defects. **g** Cross-section across the twisted plane between the two CBs. The scale bar is 5 μm

just π. This additional twist causes mismatch between the orientation of the director and the anchoring on the surface of the droplet which results in the line disclination running circumferentially around the central twisted region. Additionally the twisting of this region splits the hyperbolic region which hosted the -1 defect in the droplet with 3 point defects into two hyperbolic point defects. Effectively the central point defect from the droplet with 3 collinear defects is substituted by a combination of two hyperbolic defects and a plane of twisting director which ends with a ring disclination at the surface of the droplet. Another way to understand this structure is to compare it with the one of a ring disclination with a single cholesteric cylinder. We can construct it by pulling one of the defects at the ends of the cylinder through the disclination line and filling the space on the other side with a topologically neutral structure of a CB with a pair of $|q| = 1$ defects.

7.5.3 Other Topological Molecules

An interesting structure which can be regarded as a topological molecule was observed in a droplet shown in Fig. 7.67 at $N = 6.1$. It is not immediately obvious what the structure is from the transmission images in Fig. 7.67a–d, but the reconstructed director field in Fig. 7.67e, f shows that the structure includes the Lyre-like part from Fig. 7.14 in the bottom part of the droplet and a cholesteric bubble with an associated pair of ± 1 point defects in the top part of the droplet.

Figure 7.68 shows more detail about the structure in a series of cross-sections perpendicular to the symmetry axis of the droplet. In Fig. 7.68a we can see that Lyre part of the structure has cylindrical symmetry above the point defect and in Fig. 7.68c we can see the ring of circular streamlines, which form the core of the tube in the Lyre structure. Figure 7.68f shows that the equatorial cross-section of the -1 point defect next to the CB has a radial configuration and Fig. 7.68g shows the cylindrical symmetry of the CB.

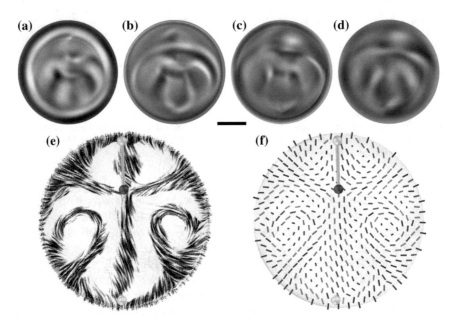

Fig. 7.67 A droplet ($N = 6.1$) with a Lyre-like structure and a cholesteric bubble. **a–d** A sequence of non-polarised images at different focusing planes separated by 4.5 μm, (**a**) being the closest to the objective. **e, f** The reconstructed structure in a plane which includes the symmetry axis and all three point defects. The locations of point defects are marked by coloured points. The scale bar is 5 μm

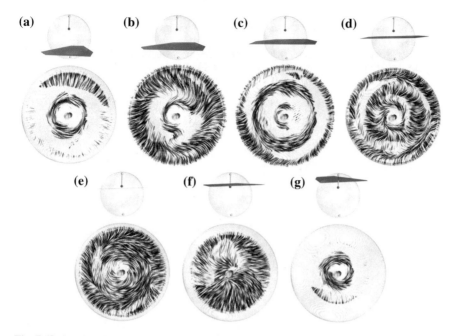

Fig. 7.68 Details of the structure of a droplet from Fig. 7.67. The streamlines show the director field in different cross-sections along the symmetry axis of the droplet with their locations indicated in the insets

References

1. G. Posnjak, S. Čopar, I. Muševič, Points, skyrmions and torons in chiral nematicdroplets. Sci. Rep. **6**, 26361 (2016)
2. I. Kézsmárki et al., Néel-type skyrmion lattice with confined orientation in the polar magnetic semiconductor GaV_4S_8. Nat. Mater. **14**, 1116–1122 (2015). http://www.nature.com/nmat/index.html
3. S. Pirkl, P. Ribiere, P. Oswald, Forming process and stability of bubble domains in dielectrically positive cholesteric liquid crystals. Liq. Cryst. **13**, 413–425 (1993)
4. D. Seč, T. Porenta, M. Ravnik, S. Žumer, Geometrical frustration of chiral ordering incholesteric droplets. Soft Matter **8**, 11982–11988 (2012)
5. S. Čopar, S. Žumer, Quaternions and hybrid nematic disclinations. Proc. R. Soc. A **469**, 20130204 (2013)
6. D. Seč, S. Čopar, S. Žumer, Topological zoo of free-standing knots in confined chiral nematic fluids. Nat. Commun. **5**, 3057 (2014)
7. G.P. Alexander, B.G.-G. Chen, E.A. Matsumoto, R.D. Kamien, Colloquium: disclination loops, point defects, and all that in nematic liquid crystals. Rev. Mod. Phys. **84**, 497–514 (2012)
8. U. Mur et al., Ray optics simulations of polarised microscopy textures in chiral nematic droplets. Liq. Cryst. **44**, 679–687 (2017)
9. G. Posnjak, S. Čopar, I. Muševič, Hidden topological constellations and polyvalent charges in chiral nematic droplets. Nat. Commun. **8**, 14594 (2017)

10. S. Čopar, S. Žumer, Topological and geometric decomposition of nematic textures. Phys. Rev. E **85**, 031701 (2012)
11. X. Wang et al., Experimental insights into the nanostructure of the cores of topological defects in liquid crystals. Phys. Rev. Lett. **116**, 147801 (2016)

Chapter 8
Schematic Construction of Droplets with Multiple Cholesteric Bubbles

In this chapter we show how we can decompose the director structures of the droplets with multiple point defects and cholesteric bubbles into basic building blocks and use them to schematically explain their structure and construct topological molecules.

8.1 Construction of String-Like Constellations

First we'll try to understand, how the strings of point defects are formed. In the droplets with low chirality, a single cholesteric bubble is present, together with a $+1$ point defect, which is needed because of the homeotropic anchoring on the surface of the droplet. This structure is shown again in Fig. 8.1d, which presents its director streamlines in a plane, including the point defect and the symmetry axis of the structure. We should keep in mind that the structure is rotationally symmetric around its axis. All of the strings with multiple point defects have an odd number of point defects, which is a consequence of homeotropic anchoring of the droplets. This means that point defects in excess of the topologically required one from the simple droplet in Fig. 8.1d, appear only in pairs. In each pair there is a -1 hyperbolic defect, a $+1$ twisted defect and a cholesteric bubble, which stabilises the distance between the two defects, so they do not annihilate.

An example of such a pair of ± 1 point defects is shown in Fig. 8.1a in a droplet with three collinear defects. We can cut the droplet into two parts along the director streamlines which lie in the mid-plane of the hyperbolic defect and have a radial configuration as seen in Fig. 7.31f. The cut should run around the hyperbolic point defect, so that the central streamlines which point along the symmetry axis of the hyperbolic point defect stay attached to the defect. In this way we have isolated a pair of ± 1 defects with one of the CBs (Fig. 8.1b) from the remainder of the droplet. What is left, is a single $+1$ point defect and a cholesteric bubble with a rift along its symmetry axis (Fig. 8.1c). Because the hyperbolic part of the streamlines has been

© Springer Nature Switzerland AG 2018
G. Posnjak, *Topological Formations in Chiral Nematic Droplets*,
Springer Theses, https://doi.org/10.1007/978-3-319-98261-8_8

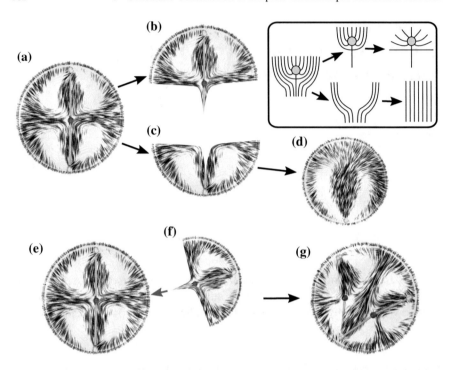

Fig. 8.1 Subtraction and addition of topological units. The droplet with 3 collinear defects in (**a**) can be divided into a pair of ±1 defects with one of the CBs (**b**) and the rest of the droplet with one +1 defect and the other CB (**c**). The remainder of the droplet in (**c**) can be smoothly transformed into a full droplet with a point defect and a CB, shown in (**d**). The inset shows an analogous procedure for a colloid-defect pair, inserted into uniform director field. If we take a droplet with three collinear defects (**e**) and insert into it a topologically neutral ±1 defect pair with a CB (**f**), we get a droplet with a V-shaped string of 5 defects, shown in (**g**). Panels a, e and g reprinted from Ref. [1] under the terms of the Creative Commons Attribution 4.0 International License (http://creativecommons.org/licenses/by/4.0/)

removed, this rift can be closed by a continuous deformation, which yields the simple droplet with a single CB and a single +1 point defect (Fig. 8.1d).

The pair of defects in Fig. 8.1b is analogous to the one which forms if a colloid with homeotropic anchoring is inserted into uniform director field. The inset in Fig. 8.1 shows how the colloid-defect pair can be separated from the homogeneous director field. Note that again the cut separates the director field along the director streamlines which originate in the equatorial plane of the hyperbolic point defect and also includes the symmetry axis of the droplet. The streamlines which originate from the equatorial plane of the hyperbolic defect can be smoothly deformed from the cone-like shape that they take when immersed in uniform bulk, to a radial orientation in a single plane as shown in the inset to Fig. 8.1. This configuration is analogous to the one in Fig. 8.1b. The pair of ±1 defects has a zero total topological charge and can therefore be inserted into uniform director without changing its topology.

The situation is similar in our droplets, where the pair of defect around a CB also carries zero total topological charge and can therefore be inserted into the radially oriented director field of droplets with total topological charge equal to $+1$.

We can test this idea by reconstructing the structure with three collinear point defects from the wedge-like half-droplet with the pair of defects and the CB in Fig. 8.1b, and the simple droplet from Fig. 8.1d. We make a cut into the mostly radial part of droplet in Fig. 8.1d, opposite to the point defect and insert the wedge of the half-droplet. The director on the outer surface of the half-droplet is mostly radial, so it matches the surface it supplanted, and the inserted structure carries zero topological charge, so we did not change the topology of the droplet.

This kind of addition of topologically neutral units can be taken further to construct more complex structures. We can start with a droplet with three collinear defects and two CBs (Fig. 8.1e), find a nice patch of radially oriented director on its side and implant the topologically neutral half-droplet from Fig. 8.1f into it. The reasoning why we can do this is similar as in the previous case—the radial surface of the droplet is basically uniform director field into which we insert the topologically neutral half-droplet. In this way we have constructed a droplet with a V-shaped string of 5 point defects and three cholesteric bubbles.

The radial director on the surface of a homeotropic droplet offers plenty of space to insert further topologically neutral units. If another pair of ± 1 defects with a CB is inserted into the droplet with 5 point defects, we get a string of 7 defects. Because such a structure has 4 surface $+1$ defects, the volume of the droplet is occupied most efficiently if the new pair of defects is not located in the plane of the previous 5. With one more insertion of a topologically neutral pair of defects, we would construct a droplet with a string of 9 defects.

The idea of modularity of these structures is reinforced by the fact that for every hyperbolic point defect in the string-like constellations we can find a plane with an almost radial orientation of director field, which shows where the additional cholesteric bubble is attached to the rest of the director structure. Examples of such planes in structures with 5, 7 and 9 point defects are shown in Figs. 7.36b, 7.40 and 7.44d, i, respectively. The radial director profiles in these examples are elastically deformed in a similar way as shown in the inset to Fig. 8.1 to match the confinement of the surrounding structure. Because of this the director in these cross-sections is not completely radial in a flat plane as in the droplet with 3 collinear point defects, but we can still recognise the radial orientation of the streamlines around the hyperbolic point defect.

8.2 Construction of Topological Molecules with Higher-Charge Point Defects

When presenting the structures of the complex topological molecules in Sect. 7.5 we helped ourselves to understand the structures by suggesting that single point defects with their associated cholesteric bubbles are interchanged by bigger structural units

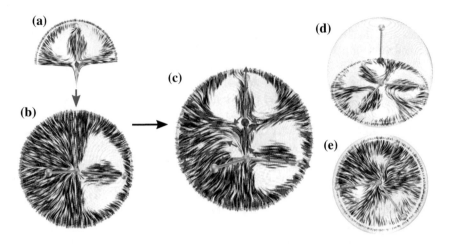

Fig. 8.2 Addition of a neutral ± 1 defect pair to a droplet with a -2 defect. The topologically neutral half-droplet (**a**), which we isolated in Fig. 8.1 is inserted from the top into a droplet with one -2 and three $+1$ point defects along with three CBs, shown in (**b**). The plane of the defects in (**b**) is horizontal and the streamlines are shown in a cross-section going through the -2 defect and one of the $+1$ defects. **c** After the neutral half-droplet is inserted, a more complex structure is formed. The red arrows indicate the direction assigned to the director to work out the signs of topological charges of the defects. **d** A different view of the resulting structure, with the streamlines shown in a plane going through the -2 defect and the surrounding $+1$ defects. **e** Streamlines in the mid-plane of the -1 defect

with the same total topological charge. This idea did not explain exactly how the inserted structure integrates into the existing one. Here we present an alternative, detailed way to explain the structures in the topological molecules by building on the idea from the previous Section of inserting topologically neutral building blocks into stable droplets.

Let's start with a droplet with a -2 defect inside a triangular arrangement of $+1$ defects from Fig. 7.47. Figure 8.2b shows the director streamlines of this droplet in a plane going through the central -2 defect and one of the $+1$ defects. The top and bottom of the droplet in Fig. 8.2b have a mostly radial orientation of director field which will serve as uniform director, just like in the simpler droplets in the previous Section. Here we can insert the topologically neutral pair of ± 1 defects with a CB shown in Fig. 8.2a without changing the total topological charge of the droplet. We can see the resulting structure in Fig. 8.2c, d, where it is obvious that the top CB retains its shape, as do also the three CBs connected directly to the -2 defect. The structure we have thus constructed matches the topological molecule from Figs. 7.58 and 7.59.

Figure 8.2e shows the equatorial cross-section of the -1 defect where the two structures from Fig. 8.2a, b are joined together. We can see that it has a radial configuration of director field, making it equivalent to the radial planes in Fig. 8.1 and in the string-like structures. Arrows are added to Fig. 8.2c to check the signs of topological charge of the point defects. We can see that the added hyperbolic defect really has a -1 topological charge, even though it is directly connected to the -2 defect. This differs from the sign-alternating behaviour of neighbouring point defects we observed in the previous structures. The reason for this is that the hyperbolic defect is not attached to one of the patches of the -2 defect like the $+1$ defects which surround it, but instead to the uniform background with the reversed direction of director field (the blue areas in Fig. 7.54a, b).

In the example in Fig. 8.2, the structure of a droplet with a -2 defect is expanded by inserting into it a pair of ± 1 defects, which is topologically neutral. In some of the other previously presented topological molecules, the neutral structure includes a higher-charge point defect. Such a complex topologically neutral building block can be obtained by starting with the simplest droplet with a -2 point defect shown in Fig. 8.3a. We cut the droplet along the hyperbolic director streamlines of one of the patches of the -2 defect in an analogous fashion as we did in the droplet with three point defects in the previous Section. In this way we separate $2/3$ of the droplet which include two $+1$ defects, two CBs and the hyperbolic region of the -2 defect (Fig. 8.3b) from the $1/3$ of the droplet which includes a single $+1$ defect and a CB with a conical hole along its symmetry axis (Fig. 8.3c). The third of the droplet in Fig. 8.3c can again be smoothly deformed to close the hole along the symmetry

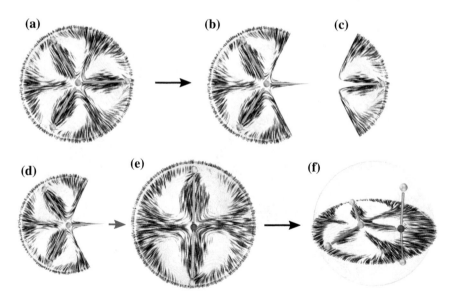

Fig. 8.3 Topologically neutral unit with a -2 point defect. A droplet with one -2 and three $+1$ point defects and three CBs, shown in (**a**), can be split into a part with a single $+1$ point defect and a CB (**c**) and the remainder of the droplet (**b**) with zero total topological charge. The topologically neutral part of the droplet with a -2 defect in (**d**) can be inserted into a droplet with a string of three point defects in (**e**) to form a complex molecule, shown in (**f**). Panels **a**, **e** and **f** reprinted from ref. [1] under the terms of the Creative Commons Attribution 4.0 International License (http://creativecommons.org/licenses/by/4.0/)

axis to get a whole droplet with a point defect and a CB, just as in Fig. 8.1d. The part of the droplet with the -2 defect also includes two $+1$ defects and therefore its total topological charge is zero. The director orientation along the surface of the cut radially spreads out from the -2 defect towards the edge of the droplet, just like it did in the case of the -1 defect in the previous Section. In this case the cut does not lie in a plane, but instead on a conically shaped surface. Nonetheless the radial orientation of the director on this surface can be seen for sections of it in Fig. 7.49a, d, f. This surface with the radial orientation of the director field will again serve as the contact area between the two structures which we will join in a molecule.

Now we can insert our topologically neutral unit with a -2 defect, which is shown again in Fig. 8.3d, into a droplet with three point defects and two cholesteric bubbles, shown in Fig. 8.3e. The radial director field on the side of the droplet in Fig. 8.3e will serve as the uniform director into which the neutral structure is implanted. After the director field is elastically relaxed in the structure, we can still recognise the outlines of the added topologically neutral part in the left side of Fig. 8.3f and the string of three unit-charge point defects, which is perpendicular to the plane of the streamlines.

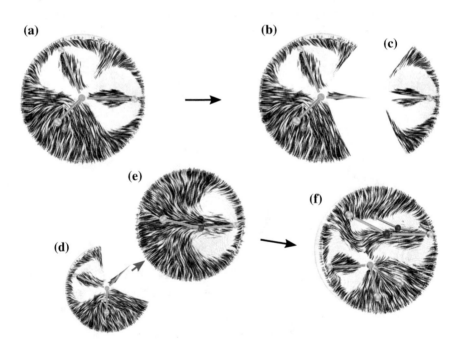

Fig. 8.4 Topologically neutral unit with a -3 point defect. A droplet with one -3 and four $+1$ point defects and four CBs, shown in (**a**), can be split into a part with a single $+1$ point defect and a CB (**c**) and the remainder of the droplet (**b**) with zero total topological charge. The topologically neutral part of the droplet with a -3 defect in (**d**) can be inserted into a droplet with a V-shaped string of 5 point defects in (**e**) to form a complex molecule, shown in (**f**). Panels **a** and **f** reprinted from Ref. [1] under the terms of the Creative Commons Attribution 4.0 International License (http:// creativecommons.org/licenses/by/4.0/)

Number of +1 defects	String	Topological molecules and higher point defects

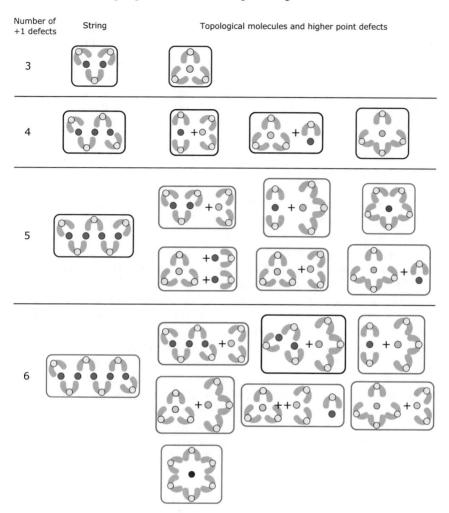

Fig. 8.5 Examples of the possible combinations of topological point defects in droplets with different numbers of +1 point defects close to the surface. For three and four +1 point defects all the possible combinations are shown, and for five and six +1 point defects only some of them. The structures which were not observed experimentally are marked by a red frame. For droplets with five and six +1 point defects, the simplest structures with the theoretically possible −4 (blue circle) and −5 (brown circle) point defects are included. These defects can be constructed by inserting one or two additional hyperbolic patches into the structure of a −3 point defect in an analogous fashion as presented in Sect. 7.4.3

The resultant structure is obviously the one from the droplet in Figs. 7.55, 7.56 and 7.57.

A similar topologically neutral building block can be extracted from a droplet with a −3 defect and four sets of +1 point defects with cholesteric bubbles, shown in Fig. 8.4a. Again we make the cut on one side of the −3 defect following the hyperbolic streamlines of the director field as shown in Fig. 8.4b to remove a quarter of the droplet with a single $+1$ point defect and a CB (Fig. 8.4c). This removed part can, as in the previous cases, be smoothly transformed to make a complete simple chiral nematic droplet with a CB. The rest of the droplet includes the whole -3 point defect along with three $+1$ defects and three CBs and therefore has a zero total topological charge. The director on the conical surface of the cut in Fig. 8.4b again has a radial configuration and will serve as the interface between the two mated structures.

We insert the constructed topologically neutral structure with a -3 point defect, shown again in Fig. 8.4d, into a droplet with a V-shaped string of 5 point defects (Fig. 8.4e). The radially oriented director field on the left part of the droplet in Fig. 8.4e serves as the uniform director field into which we insert the neutral structure and when the director is elastically relaxed we obtain the structure in Fig. 8.4f, which was presented in detail in Figs. 7.60, 7.61 and 7.62.

By using the idea of inserting complex but topologically neutral structures into stable chiral nematic droplets, we can also construct topological molecules, which we did not observe in experiments. Some of the possible combinations of topological charges for a given number of $+1$ point defects are shown in Fig. 8.5. The structures in Fig. 8.5 are classified with respect to the number of $+1$ point defects close to the surface of the droplet. The ones which we observed experimentally are enclosed in black frames, and others which are schematically possible in red frames. We can see that all the possible combinations for three and four $+1$ point defects were discovered. For both five and six $+1$ point defects only one type of a structure was found. No systematic search for the missing structures has been performed yet so perhaps their absence is only a matter of unfavourable statistics, but it could also indicate that the deformations which would be present in them are energetically not favourable. A more thorough investigation could lead to insights into the problem of packing of the elastic cholesteric bubbles into a spherical volume.

Reference

1. G. Posnjak, S. Čopar, I. Muševič, Hidden topological constellations and polyvalent charges in chiral nematic droplets. Nat. Commun. **8**, 14594 (2017)

Chapter 9
Discussion

In this Thesis, we have presented a vast array of structures formed in chiral nematic droplets with homeotropic anchoring. At low relative chiralities ($N = 2d/p_0 \approx 2 - 4$), the topologically necessary point defect is expelled from the centre of the droplet and the bulk is filled with a chiral structure shown in Fig. 9.1a, which we call a cholesteric bubble. At higher chiralities, by far the most common structure is one with a single point defect near the surface of the droplet and bent cholesteric layers in the bulk as shown in Fig. 9.1b. The number of cholesteric layers in these structures increases with the relative chirality of the droplet. Many other director configurations are possible over a wide range of values of relative chirality. An overview of the structures, presented in this Thesis, is shown in Figs. 9.1 and 9.2, together with a graph of the chirality ranges at which they were observed.

We can divide the structures roughly into two subcategories: droplets with layered structures shown in Fig. 9.1b–o which also include droplets with ring disclinations (Fig. 9.1k–o), and droplets with several cholesteric bubbles shown in Fig. 9.2, which include the higher-charge point defects (Fig. 9.2e, h) and their topological molecules (Fig. 9.2f, g, i). The different structures in this cholesteric zoo appear in overlapping ranges of relative chirality values shown in Fig. 9.2j. The structures with bent or flat layers in Fig. 9.1b, c are excluded from the graph because their layeredness can be continuously varied and it is difficult to classify them in a unique way. We can see that most of the structures can be found over a wide range of relative chiralities. This means that a specific shape can elastically deform or "resize" to fit into droplets with different relative chiralities. From this we can speculate that a structure has different elastic energy densities for different relative chiralities, with an energy minimum somewhere in the middle of its stability range. On the other hand, for any given relative chirality in the range $N = 2.5 - 6.8$ several different structures with different elastic distortions and different types and numbers of defects are possible. All of the observed structures are stable on a time scale of several days, which means that each structure is at least a local minimum of the free energy and thus metastable.

© Springer Nature Switzerland AG 2018
G. Posnjak, *Topological Formations in Chiral Nematic Droplets*,
Springer Theses, https://doi.org/10.1007/978-3-319-98261-8_9

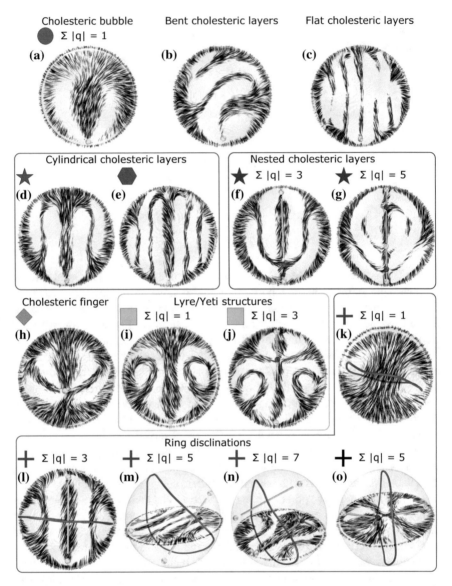

Fig. 9.1 An overview of the presented structures in chiral nematic droplets with homeotropic anchoring, Part I. The structures are grouped with like structures. The symbols next to the images of the streamlines are the symbols used in the graph in Fig. 9.2j to present the stability ranges of the structures. Where the use of the symbols in the graph could be ambiguous, the sum of the absolute value of the topological charges in the structure is given. Line defects are considered as having $|q| = 1$, even though they could have a higher odd topological charge

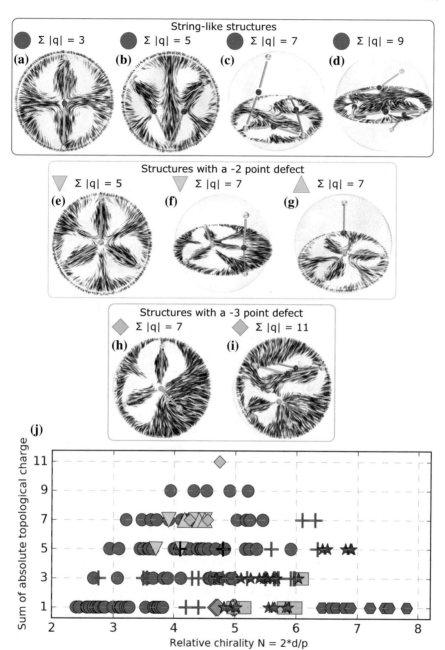

Fig. 9.2 An overview of the presented structures in chiral nematic droplets with homeotropic anchoring, Part II. A graph of the stability ranges is given in panel (**j**)

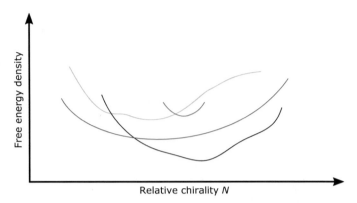

Fig. 9.3 A schematic example of the possible free energy density dependence of different stable structures with respect to the relative chirality N. Each coloured curve belongs to a different structure. Intersections between the energy curves do not imply transitions between the structures because of energy barriers associated with deformations of the structures which are needed for the transitions

We can try to illustrate this by a schematic graph of a free energy dependency on the relative chirality N, shown in Fig. 9.3, where the free energy densities of different structures are shown with curves of different colours. Each structure is stable over a range of N values but there is some ideal N, at which the different types of elastic deformations in a structure have the lowest combined free energy and the structure is most stable. At a fixed N value, different structures have different free energies, as shown in Fig. 9.3. Intersections of the free energy curves in Fig. 9.3 do not imply that spontaneous transitions between the states are possible. For transitions between states, the structures have to be deformed or even locally melted, which means that there are energy barriers associated with these transitions. The energy barriers have to be considerably higher than the energy of thermal fluctuations, because of the stability of the presented structures.

As an exercise we can try to imagine how two different structures can be transformed between each other in a hypothetical scenario, which does not imply these transitions are energetically favourable. A transition between two states can be of two types: either a continuous transformation, where we can transform one structure into another through a series of finite deformations, or a discontinuous transformation, where the cholesteric layers have to be cut and rewired. An example of a smooth, continuous transformation is shown in Fig. 9.4a, b, where the ring and cholesteric cylinder structure transitions into the cholesteric bubble and cholesteric cylinder structure. The ring defect in Fig. 9.4a can slide along the surface towards one of the point defects as indicated by the red arrows. As the ring defect shrinks into a point defect, the layers of twist next to it reshape to form a cholesteric bubble and push the nearby point defect away from the surface and through the defect loop as indicated by the blue arrow. The smoothness of this transformation implies that the two structures are topologically related or almost equivalent, which is true if we consider the point defect to be a small defect ring, similarly as in the case of a hyperbolic defect

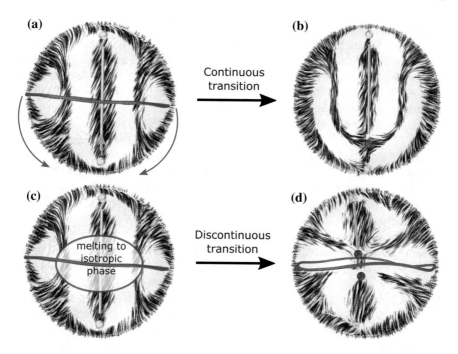

Fig. 9.4 Two examples of hypothetical transformations between the observed structures in chiral nematic droplets with homeotropic anchoring. **a, b** A smooth, continuous transformation between (**a**) the ring and cholesteric cylinder structure with two +1 point defects close to the surface and (**b**) the structure of a cholesteric cylinder, nested inside a cholesteric bubble. The red arrows indicate the movement of the ring defect and the blue arrow of the bottom point defect, during the transformation. **c, d** A discontinuous transition from the ring defect and cholesteric cylinder structure in (**c**) to the ring defect and two cholesteric bubbles structure in (**d**). The central part of the droplet in (**c**) has to be melted to isotropic phase to cut the cholesteric cylinder and rewire the cholesteric layers, so two cholesteric bubbles and two hyperbolic defects (purple points in (**d**)) can be formed

next to a particle with homeotropic anchoring, which can open into a loop [1, 2]. This interpretation of a point defect agrees with theoretical calculations on the form of the defect core [3], which was recently demonstrated experimentally by loop-like formation of amphiphilic molecules, polymerised in the core of a +1 point defect [4].

A possible example of a discontinuous transformation is from the ring and cholesteric cylinder droplet in Fig. 9.4c to the ring and two cholesteric bubbles structure in Fig. 9.4d. In this case, the centre of the droplet in Fig. 9.4c has to be melted to cut the cylinder in half and to enable rewiring of the cholesteric layers. Two hyperbolic defects have to form after the cut to stabilise the two halves of the cholesteric cylinder, which are reshaped into two cholesteric bubbles.

More information about the free energy landscape of the structures can be extracted from the frequency of their appearance. Most of the structures can form spontaneously while mixing the liquid crystal with the carrying medium at room temperature. The most common structures are the cholesteric bubble (Fig. 9.1a) and the bent cholesteric layers structures (Fig. 9.1b), which are by an order of magnitude more frequent than the other configurations in their respective stability ranges. Out of the other structures, the most common ones are the linear string of 3 point defects with two cholesteric bubbles (Fig. 9.2a), droplets with flat cholesteric layers (Fig. 9.1c) and the ring defect and cholesteric cylinder structure (Fig. 9.1). All of the cholesterically layered and cylindrically symmetric structures appear in unquenched samples, with the exception of the cholesteric finger (Fig. 9.1h) and the Lyre/Yeti structures (Fig. 9.1i, j). The only string-like constellation beside the linear string of 3 point defects we observe in unquenched samples is the V-shaped string of 5 point defects (Fig. 9.2b). In general, a temperature quench increases the probability of formation of the more complex structures, but all of the structures from unquenched samples are still possible. The most complex structures like the ones with double cholesteric cylinders, higher-charge point defects and the topological molecules are formed only after temperature quenches.

There are two main factors which need to be taken into account when considering the frequency of appearance of the different structures. The first and the most obvious one is the free energy of the structure, which might be expected to be relatively low for the most common structures and comparatively high for the rarer ones. The second factor that affects the probability of structure formation is how specific are the starting configurations of the director which can relax to a certain structure. For example, mixing the LC into the carrying medium only slightly perturbs the cholesterically layered ground state of the chiral nematic, which results in layered and cylindrically symmetric structures being more common in unquenched samples. On the other hand, a quench induces a random starting configuration which can relax into any of the possible structures which are local minimums of the free energy and allows formation of the more complicated structures with many structural features such as the different defects, cholesteric bubbles and cholesteric cylinders. This suggests that complex structures do not necessarily have higher free energy than the more common ones, but that they might occupy a part of the configuration space which can only be reached by specific starting conditions which almost never arise if the droplets are formed by mixing, and only rarely if the sample is quenched. This makes the complex states difficult to access and in a sense, hidden, similarly to the recently found hidden quantum states in electronic systems [5]. In this study, these states are accessed randomly and consequently the frequency of their appearance is low. A way to ensure their formation would be to somehow seed the configuration of the director field after the quench, i.e. induce a favourable starting configuration with an external field, for example with a structured laser beam [6].

It is interesting to compare the observed structures with the numerically predicted ones. In bulk volumes, chiral nematic liquid crystals are expected to form cholesteric layers because of the helical twisting of the director and like-wise, layeredness is a prominent feature of most of the known structures in chiral nematic droplets. A numerical study by Seč et al. [7] predicted the formation of extended line defects where the cholesteric layers meet the homeotropic surface of the droplet, as shown in Fig. 9.5a, b. The most common structure we observe in our samples (Fig. 9.5c) really does have a cholesteric layered structure, but its cholesteric layers are bent

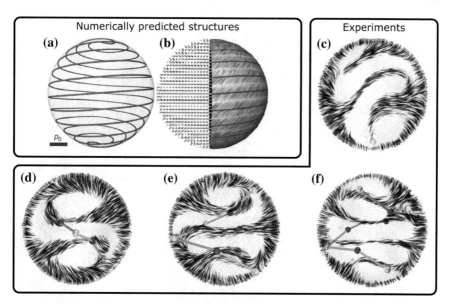

Fig. 9.5 Comparison of numerically predicted and experimentally observed layered structures in chiral nematic droplets with homeotropic anchoring. The numerical droplet in **a, b** at $N = 12$ has a singular defect line spiralling along its surface where the cholesteric layers meet the homeotropic surface as shown in (**a**) and in (**b**) we can see the cholesteric layers in its structure. In the experimental structure with bent layers in **c** all the topological charge is concentrated in a single point defect and the mismatch between the layers and the homeotropic surface is solved by an escaped structure. **d–f** Experimental structures with strings of 5, 7 and 9 point defects. The bulk of the droplets has a layered structure and the cholesteric bubbles in the string couple the layers to the homeotropic surface. Panels **a** and **b** reprinted by permission from Macmillan Publishers Ltd: Nature Communications [7], copyright 2014

and no surface disclination lines are present. This is because the director orientation mismatch is resolved in an escape-like structure in areas where the cholesteric layers reach the surface. The deformation in this region is a mix of all three geometrical deformations. It is difficult to classify this escape as a cholesteric defect, as the proper cholesteric defects which are presented in Sect. 2.5.3 are surrounded by a cholesteric medium and in this case one side of the defect is the homeotropic surface, which is non-chiral because of the constraint $\mathbf{n} \cdot (\nabla \times \mathbf{n}) = 0$ [8]. These escaped regions do not carry any topological charge, as the single present topological point defect satisfies the restraint on topological charge. Droplets with linear strings of point defects shown in Fig. 9.5d–f also have a layered structure in the bulk, but here the strings of point defects and the associated cholesteric bubbles which run around the droplet serve as a part of the interface between the cholesteric layers in the bulk LC volume and the non-chiral surface.

The cylindrically symmetric layered structures which we observe in our experiments show striking similarity to structures which are numerically predicted for chiral nematic droplets with planar anchoring [9]. Figure 9.6 shows a comparison

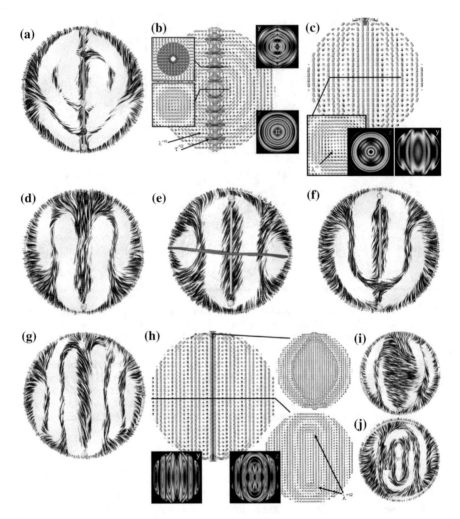

Fig. 9.6 Comparison of experimental structures in droplets with homeotropic anchoring (**a, d–g, i, j**) and numerically predicted ones in droplets with planar anchoring (**b, c, h**). The nested structure with 5 point defects in (**a**) is related to the diametric spherical structure in (**b**). The bipolar structure in (**c**) has three related experimental structures: **d** the structure with cylindrical layers, **e** with a ring defect and a cholesteric cylinder, and **f** the cholesteric cylinder, nested in a cholesteric bubble. The experimental structure with flat cholesteric layers in (**g**) is related to the planar bipolar structure in (**h**) which is confirmed by comparing additional cross-sections in (**i, j**). Panels **b, c** and **h** are reproduced from Ref. [9] with permission of The Royal Society of Chemistry

of the structures, where the experimentally found ones are shown in black director streamlines and the numerical ones with blue cylinders. The bulk of the nested structure in Fig. 9.6a is an onion-like structure with concentric layers, similar to the diametric spherical structure in Fig. 9.6b. In the homeotropic case, a string of point defects runs diametrically across the structure, with each point defect being positioned between two cholesteric layers or a layer and the surface of the droplet. The numerical study from Ref. [9] found that in the planar case, the layers are separated by small $\tau^{-1/2}$ disclination rings with singular cores. These ring defects carry topological charge which makes them equivalent to topological point defects. The main difference between the two types of structures is, that in the diametric spherical structure the diametric string of defects ends in two surface boojums, whereas in the homeotropic case there are no surface boojums, but the onion-like layered structure is nested in a cholesteric bubble with an associated point defect, which satisfies the topological restrictions of the homeotropic anchoring.

Figure 9.6c shows the cylindrically symmetric bipolar structure from the numerical study by Seč et. al. [9] which has three analogue homeotropic structures shown in Fig. 9.6d–f. All four structures have coaxial cylindrical layers and no defects along their symmetry axes except close to the surface, where the planar structure has a pair of boojums and the homeotropic structures have either one or two point defects. The structure in Fig. 9.6e has an additional circumferential ring defect where the outer layer of twist reaches the homeotropic surface of the droplet.

The homeotropic structure with flat layers in Fig. 9.6g is similar to the planar bipolar structure from Ref. [9], shown in Fig. 9.6h. We can see from the two additional cross-sections of the homeotropic structure in Fig. 9.6i, j that the structures are indeed almost identical apart from the positioning of defects: the homeotropic structure has a single bulk point defect near the surface and the planar bipolar structure has two diametrically positioned surface boojums.

In Fig. 9.7 the Lyre/Yeti structure (Fig. 9.7a) in homeotropic droplets is compared to the Lyre (Fig. 9.7b) and Yeti (Fig. 9.7c) structures from Ref. [9]. All three structures are cylindrically symmetric and the homeotropic Lyre/Yeti structure combines features of both planar structures.

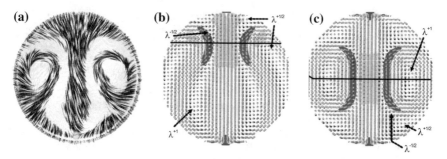

Fig. 9.7 Comparison of **a** the experimental Lyre/Yeti structure in homeotropic droplets and the numerically predicted **b** Lyre and **c** Yeti structures in planar droplets. Panels **b**, **c** are reproduced from Ref. [9] with permission of The Royal Society of Chemistry

All these structures in homeotropic droplets have one thing in common: the bulk of their volume has a structure, similar to the planar droplets. In a way, these homeotropic structures can be understood as the structures of the planar droplets, inserted into a shell with planar anchoring on the inner surface and homeotropic anchoring on the outer one. The boojums of the planar structure are positioned on the inner surface of the shell and the director in the shell can either transform them into proper bulk point defects, or resolve them in a non-singular way. In structures in Figs. 9.6d, g and 9.7a one of the boojums is resolved and the other becomes a single point defect close to the surface, which is topologically demanded because of the homeotropic anchoring conditions. In structures in Fig. 9.6a, e, f both of the boojums are transformed into bulk point defects and the shell includes an additional point or ring defect to satisfy the homeotropic anchoring conditions. The outer shell therefore serves to couple the planar orientation of director on the surface of the inner structure to the homeotropic anchoring on the outer surface, and to provide the topologically demanded charge.

It is a fact that we haven't observed the extended line defects, knots and links which were numerically predicted to form in quenches of chiral nematic droplets with homeotropic anchoring. This doesn't mean that the extended lines do not form in the droplets. There are considerable differences between the conditions of the experiments and the numerical quenches. For one, the numerically studied droplets are by an order of magnitude smaller than the experimental ones because of restrictions on computation time and memory. Additionally, the numerical simulations were conducted with relatively high values for anchoring strength, at a temperature relatively close to the phase transition, and with the one constant approximation for elastic deformations. All these factors could contribute to differences in the stability of the different types of structures.

An instructive example is the comparison of the theoretical and experimental structures in Fig. 9.8, where the direction of the helical axis in the cholesteric regions

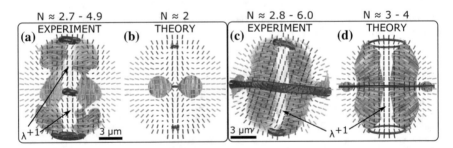

Fig. 9.8 A comparison of the cholesteric and singular regions in experimental (**a, c**) and numerically simulated (**b, d**) chiral nematic droplets with homeotropic anchoring. The structure with a linear string of 3 point defects and two cholesteric bubbles is shown in (**a, b**) and in (**c, d**) the structure with a defect ring and a cholesteric cylinder with two point defects. The streamlines show the direction of the helical axis where it can be determined and the red areas show the defects. Reprinted from Ref. [12] under the terms of the Creative Commons Attribution 4.0 International License (http://creativecommons.org/licenses/by/4.0/)

is shown in streamlines. The helical axis is defined as a direction perpendicular to the director, around which the director rotates. We follow the definition of the chiral axis from Ref. [10] to extract it from the data. Before determining the helical axis we have to smoothen the experimental data, because the calculation involves spatial derivatives of the director field. We do this by feeding the experimental reconstructed director into the code for minimisation of the Q-tensor Landau-de Gennes free energy by relaxation which is usually implemented in numerical simulations of achiral and chiral nematics [11]. The experimental director field is relaxed with several steps of numeric relaxation to reduce the experimental noise and artefacts which arise during the reconstruction. Because the Q-tensor formalism is used at this stage, the order parameter S is reduced during the relaxation in areas with large jumps in the director field. The areas of reduced S are shown in red in Fig. 9.8.

We can see that the structure with a linear string of three point defects in Fig. 9.8a, b is similar in the experimental and numerical case, but in the numerical simulation it appears at slightly lower relative chirality values and we can see that the helical axis is defined only in a small part of the volume. This is because in the numerical structure, the cholesteric bubbles do not have a Bloch-like behaviour where the director field in their cross-section twists from the central orientation like in the experimental structure in Fig. 9.8a. Instead they feature a Néel-type bending of the director from the central to the outer orientation. This might be because of the degeneracy of deformations in the one constant approximation, where the free energy of the structure is not lowered in configurations with a lot of twist.

The second set of panels in Fig. 9.8c, d shows, that the areas with defined helical axis closely match in the experimental and numerical examples of the structure with a ring defect and a single cholesteric cylinder, but there are differences in the defect regions. We can see that the defects on the bottom and top of the numerically obtained structure are opened up into defect rings, whereas in the experimental case they are shrunken into point defects. This could be because the combination of the stronger anchoring and smaller droplet size makes the material effectively stiffer in the simulations.

This comparison of experimental and numerically obtained structures suggests that with proper selection of experimental conditions, e.g. stronger confinement or stiffer material, the observed point defects could in principle be forced to open up into loops, and the director orientation mismatch between the cholesteric layers and the homeotropic surface could be resolved with singular defects instead of the observed escaped structures.

One striking feature of the observed structures is that all the higher-charge point defects along with the simpler hyperbolic ones are confined to the inner volume of the droplets, with radial and twisted radial defects being the only types which appear close to the surface. This is due to the geometrical incompatibility of the hyperbolic director structures with the uniform director field at the edge of the droplet.

Topological defects in 3D with topological charges larger than ± 1 are theoretically predicted to be unstable in achiral liquid crystals [13]. In this Thesis, higher-charge defects (Fig. 9.2e–i) are found in the confined environment of the droplets, where they are stabilised by the chirality of the liquid crystal which acts as a stabilising spring and prevents their dissociation. In Fig. 9.2j we can see, that the higher-charge structures appear in the middle of the stability ranges of string-like constellations with the same number of positive point defects, which have the same number of cholesteric bubbles. This suggests that there is an optimal confinement at which the cholesteric bubbles compress the hyperbolic defects in the bulk of the droplet in just the right way so the higher-charge defects can form. The stability of the higher-charge point defects in droplets where the symmetry of the structure does not match the valence of the defect (Fig. 9.2f, g, i), demonstrates that these defects do not exist only when the arrangement of the surrounding cholesteric bubbles exactly matches their symmetry, but are stable to modifications of the surrounding structure.

The reconstructed experimental director structures demonstrate the utility of our augmented FCPM method. The structures can be reconstructed from experimental data with very few assumptions about the observed structure. Namely, an approximate model of the elastic energy is needed and FCPM intensity values for vertical and horizontal orientation of the director field have to be determined. The finding of the missing z component sign is a numerical optimisation and the whole reconstruction procedure can be understood as an approximation of the director field which best matches the experimental FCPM data. The presented reconstructed structures exhibit artefacts, especially because of the overestimations of offset and normalisation corrections of I_{tot}. Because of this, they cannot be claimed to exactly represent the experimental structures, but in cases where the reconstructed director field is continuous, the topology of the director field is maintained, so the results are useful for topological analysis of the experiments and qualitative understanding of the structures. Numerical calculations on the experimentally reconstructed director fields are hindered by the noise and artefacts, but the examples in Fig. 9.8a, c illustrate that with suitable smoothing even quantitative analysis of the reconstructed director field is possible.

References

1. T. Lubensky, D. Pettey, N. Currier, H. Stark, Topological defects and interactions in nematic emulsions. Phys. Rev. E **57**, 610 (1998)
2. H. Stark, Physics of colloidal dispersions in nematic liquid crystals. Phys. Rep. **351**, 387–474 (2001)
3. E. Penzenstadler, H.-R. Trebin, Fine structure of point defects and soliton decay in nematic liquid crystals. J. Phys. **50**, 1027–1040 (1989)
4. X. Wang et al., Experimental insights into the nanostructure of the cores of topological defects in liquid crystals. Phys. Rev. Lett. **116**, 147801 (2016)
5. L. Stojchevska et al., Ultrafast switching to a stable hidden quantum state in an electronic crystal. Science **344**, 177–180 (2014)
6. I.I. Smalyukh, Y. Lansac, N.A. Clark, R.P. Trivedi, Three-dimensional structure and multistable optical switching of triple-twisted particle-like excitations in anisotropic fluids. Nat. Mater. **9**, 139–145 (2010)
7. D. Seč, S. Čopar, S. Žumer, Topological zoo of free-standing knots in confined chiral nematic fluids. Nat. Commun. **5**, 3057 (2014). https://www.nature.com/ncomms/
8. S. Čopar, Private communication

9. D. Seč, T. Porenta, M. Ravnik, S. Žumer, Geometrical frustration of chiral ordering in cholesteric droplets. Soft Matter **8**, 11982–11988 (2012). https://dx.doi.org/10.1039/C2SM27048J

10. D.A. Beller et al., Geometry of the cholesteric phase. Phys. Rev. X **4**, 031050 (2014)

11. M. Ravnik, G.P. Alexander, J.M. Yeomans, S. Žumer, Mesoscopic modelling of colloids in chiral nematics. Farad. Discuss. **144**, 159–169 (2010)

12. G. Posnjak, S. Čopar, I. Muševič, Points, skyrmions and torons in chiral nematic droplets. Sci. Rep. **6**, 26361 (2016)

13. H. Brezis, J.-M. Coron, E.H. Lieb, Harmonic maps with defects. Commun. Math. Phys. **107**, 649–705 (1986)

Chapter 10
Conclusion

The studied chiral nematic droplets with homeotropic anchoring show a remarkable richness of different structures. The presented structures barely scratch the surface of the possible variations. In fact, almost every time a quenched sample was put under the confocal microscope, new structures were found. This variety is even more baffling if only wide-field textures are observed. With time and experience, you learn to recognise the different orientations of the known structures, but one is quickly dumbstruck by an unknown texture, which could present a completely new structure, or just be a variation of a known one. Because of this richness of possibilities it was crucial to first develop a method which would enable us to reliably characterise the structures, regardless of their orientation. The existing state-of-the-art FCPM methods of examining director fields work wonderfully when the orientation of the structure is known or if a theoretical model of the structure exists, so that the experiment and the model can be compared. With droplets we don't have such luxury and therefore an even stronger method was needed which could characterise them solely from experimental data. The augmented FCPM method which is developed in this Thesis, achieves this remarkably. We did not discover the knotted defects of the numerical study which motivated the development of the method, but we have found much more.

The augmented FCPM method enabled us to systematically study the structures of chiral nematic droplets with homeotropic anchoring. We have found layered cholesteric structures, with either flat, cylindrical or spherical layers, which are similar to the structures in chiral nematic droplets with planar anchoring, but have a border region which includes a point defect and masks the outside homeotropic surface from the interior of the droplet. While line defects are possible and also quite common in some of the structures, point defects are the predominant form of regions with singular director. They are stabilised by localised twisted regions either in the form of cholesteric bubbles, each binding a single $+1$ point defect or in the form of

© Springer Nature Switzerland AG 2018
G. Posnjak, *Topological Formations in Chiral Nematic Droplets*,
Springer Theses, https://doi.org/10.1007/978-3-319-98261-8_10

long cholesteric cylinders with a point defect at each end. The cholesteric bubbles and their associated $+1$ point defects usually appear in pairs with hyperbolic -1 point defects, with which they form topologically neutral pairs, stable in the confinement of droplets. These pairs can be "inserted" into the droplets to form strings of odd numbers of ±1 defects.

At certain N values, the cholesteric bubbles can arrange in symmetric formations, which induce higher topological charge defects. In this Thesis we present two types of such defects with -2 and -3 topological charge. These defects are generalisations of the hyperbolic -1 defect, with an increased number of "patches". Unlike the -1 hyperbolic defect which has two patches, the higher defects are in contact with 3 or 4 cholesteric bubbles, making then, in a way, polyvalent. This polyvalency enables the formation of complex topological molecules, where one pair of a $+1$ defect and a cholesteric bubble is replaced with a bigger structure of several point defects and extended twisted regions, with a total topological charge equal to $+1$. This leads to several possible structures in droplets with tetrahedral orientation of the surface $+1$ defects and also to droplets with an octahedral structure. A peculiar feature of the droplets with point defects is that all the hyperbolic defects are expelled from the surface into the bulk volume of the droplets. We attribute this to the geometrical incompatibility of the hyperbolic shape with the surface anchoring condition, but deeper topological reasons may be lurking in the background. A handful of other exotic structures with topologically non-trivial non-singular (except for the necessary point defect, which is dictated by the confinement) director fields are presented to illustrate the richness of the structures which can appear in this simple system.

The discovered structures have fundamental importance. The higher charge point defects are a new type of point defects which have not been yet observed in any other system. Also, the arrays of localised twisted structures in the droplets, either in the form of strings or molecules, show new possible stable structures, which could have analogues in other chiral vector fields. On the other hand, the rich variety of possible metastable structures could be useful for applications, for example for memory storage [1, 2] or in optical applications as polarisation state transformers [3]. The surface $+1$ defects in regular triangular and tetrahedral arrangements could be infiltrated with nanoparticles coated with long polymer chains, which could serve as contact points for controlled assembly, for example as realisations of the tetravalent colloids [4].

Many questions remain open, for example the detailed structure of the defect core in the higher charge topological defects, structural transitions between different metastable structures, and controlled switching between the states. Furthermore, we have not found the knotted defect structures, which were predicted in the numerical quenches. Partial matching of experimentally discovered structures and some of the numerical results gives credence to their validity. Perhaps also the knotted structures can be realised experimentally with a suitable choice of parameters. The numerical quenches were conducted in by an order of magnitude smaller droplets compared to the experimental ones and the elastic constants and anchoring strength were relatively high, making the material quite stiff. A possible approach for realisation of the knotted defect lines would therefore be to use materials with elastic constants closer to the numerical parameters.

Alternatively, we could search for the knotted defects in smaller droplets with shorter pitch. Smaller droplets have smaller structural details, which are difficult to resolve using conventional optical microscopy. Even the droplets in the 10 to 20 μm range we present in this Thesis, test the state-of-the-art observation and analysis of vector fields in 3D, and indeed the current methods needed to be expanded to achieve our goals. Perhaps for smaller structures a different experimental approach would be more suitable, for example scattering methods, where the topological properties of the LC structure could be imprinted in beams of light propagating through the structure. Another option would be taking multiple projections of the structure at different orientations and connecting the image textures with the possible projections of the structure. This would probe the director structure indirectly as it would not present information about director orientation but only its effects on the probing light. Analysis of such data would demand an even more involved theoretical approach with inputs from topology and knot theory.

With the study of chiral nematic droplets we have demonstrated the effectiveness and strength of our extended FCPM method. The experimental acquisition of multiple projections of a director field in conjunction with the simulated annealing algorithm enabled us for the first time to fully reconstruct complex three-dimensional director fields from experimental data. We demonstrated and characterised the effectiveness of the method on numerical structures of complex chiral nematic droplets and we successfully implemented the method on experimental data, which enabled us to conduct the detailed study of chiral nematic droplets. We also tested the method on the well-known structures in the bubble domain texture, also called torons. The method implemented in this study uses the chiral nematic elastic energy expression, but there is no obvious reason why another energy function could not be used. An open challenge for the method is to implement it on other LC systems, e.g. complex non-chiral nematic structures, smectics, active systems, etc. The augmented method can of course be used with any other modality of confocal microscopy, which is suitable for FCPM measurements. An interesting option would be to use it with super-resolution microscopy methods, for example STED microscopy. The increased resolution of these methods could enable detailed studies of structures in blue and twist grain boundary phases.

There are no obvious reasons why our extension of the FCPM method could not be implemented in the analysis of data outside the field of liquid crystals and microscopy. In principle the simulated annealing algorithm could be used to deter-mine the experimentally missing components of a vector field in any situation where only projections of the vector field are available, for example in studies of 3D velocity fields. This kind of experimental approach, augmented with numerical optimisation extends the reach of experimental methods.

References

1. I. Muševič, S. Žumer, Liquid crystals: maximizing memory. Nat. Mater. **10**, 266–268 (2011)
2. F. Serra, M. Buscaglia, T. Bellini, The emergence of memory in liquid crystals. Materials Today **14**, 488–494 (2011)
3. E. Brasselet, N. Murazawa, H. Misawa, S. Juodkazis, Optical vortices from liquid crystal droplets. Phys. Rev. Lett. **103**, 103903 (2009)
4. D.R. Nelson, Toward a tetravalent chemistry of colloids. Nano Lett. **2**, 1125–1129 (2002)

Printed in the United States
By Bookmasters